AF519606

ACREAGE RESPONSE OF OILSEED CROPS IN A BACKWARD REGION

ACREAGE RESPONSE OF OILSEED CROPS IN A BACKWARD REGION

By

B. D. ENGALE

Reader and Head, Department of Economics
Sharadchandra College, Naigaon,
Dist. Nanded (M.S.)-431709

V. B. BHISE

Professor and Head, Department of Economics
Dr. Babasaheb Ambedkar Marathwada University,
Aurangabad (M.S.)-431004

SERIALS PUBLICATIONS
NEW DELHI (INDIA)

© B. D. Engale and V. B. Bhise

First Published - 2009

ISBN: 978-81-8387-245-4

All rights reserved with the Publisher, including the right to translate or reproduce this book or parts thereof except for brief quotations in critical articles or reviews.

Published by

SERIALS PUBLICATIONS
4830/24, Prahlad Street, Ansari Road
Darya Ganj, New Delhi-110002 (India)
Phone : 23245225 Fax : 91-11-23272135
E-mail: serials@satyam.net.in

Contents

Preface

India is a developing country, where the population has been expanding very rapidly and the levels of income are rising. The demand for agriculture products becomes more important and that's why growth of agricultural sector is essential. Indian economy is still an agricultural economy wherein about 70 percent population is dependent on agriculture. Agriculture is an important sector of the economic activity and it plays an important role in the economic development of the country.

To increase the agriculture production one should know the important factors to which it responds more. Agriculture production is a complex process. It depends upon several factors like weather, rainfall, irrigation, cropping pattern, technology, availability of high yielding variety of seeds, fertilizers, insecticides, pesticides, animal labour, farm machinery and implements and farm prices.

India holds an important position in the world in terms of rich diversity of oilseed crops and in terms of area under it. India being a major producer of oilseeds has about 23 million hectares of land under oilseed cultivation. India accounts for 10 percent of oilseed production for which uses 20 pre cent of world land under oilseeds. Oilseeds constitute the second largest crop after foodgrains in India. Oilseeds together account for 10.14 per cent of the gross cropped area.

The present book attempts to probe in to the factors which are important in determining acreage of oilseed crops, growth in area ,yield and production of oilseed crops, the area response of important oilseed crops in Marathwada region. This book also makes important suggestions to increase acreage and production of oilseed crops in the Marathwada region.

We are thankful to all concerned for providing us with the required statistical information. We acknowledge with thanks the co-operaton received from the librarian and staff of the Dr. Babasaheb Ambedkar Marathwada University, Aurangabad, Marathwada Agricultural University, Parbhani, Gokhale Institute of Politics and Economics, Pune, Swami Ramanand Teerth Marathwada University, Nanded.

Lastly, we are thankful to Mrs. Kher, Mrs. Chandrachud, Mr. Muraleedharan and Mr. V. Panchal for the long hours of typing of this manuscript and large number of tables contained in it and data processing within a short period of time.

B. D. Engale
V. B. Bhise

1 Introduction

IMPORTANCE OF SUPPLY RESPONSE STUDIES

India is a developing country, where the population has been expanding very rapidly and the levels of income are rising. The demand for agriculture products becomes more important that's why growth of agricultural sector is essential. Analysis of the price responsiveness of agricultural supply, particularly in the developing country like India, where agricultural supply responds readily to price changes in production can be accomplished through relative changes in market conditions. The price factor is thus, one of the most important factors affecting agricultural production. Agriculture plays an important role in the Indian economy. Indian economy is still an agricultural economy wherein about 70 per cent population is dependent on agriculture. Agriculture is an important sector of the economic activity and it plays an important role in the economic development of the country.

Agriculture sector provides raw materials to the industrial sector and creates demand for industrial products. Foreign currency is also obtained by exporting agriculture commodities. So, agricultural sector still plays an important role in the process of economic development of the country. So efforts are to be made to increase the production of agriculture. To increase the agriculture production one should know the important factors to which it responds more. Agriculture production is a complex process. It depends upon several factors like weather, rainfall, irrigation, cropping pattern, technology, availability of high yielding variety of seeds, fertilizers, insecticides, pesticides, animal labour, farm machinery and implements, input-output ratio and farm prices. The price is one of the most important factors affecting agricultural production on the one hand and agricultural economy on the other hand.

Apart from price factor non-price factors are also important. More precisely, price and non-price factors are jointly effective and influence in stimulating agricultural production. Therefore, both price and non-price stimulations should be provided for the agricultural production.

But two opposite views have been observed regarding responsiveness of agriculture to price factors. One view held by Dalton and others is that the farmers in the underdeveloped countries need only a small amount of cash. It was held by Mathur and Ezekeil that farmers respond negatively to changes in prices as they can get the required fixed amount of cash by selling less and keep more for their own consumption. However, some of the economists are of the opinion that price factor plays an important role in stimulating agricultural production. Price expectations in a commercialized agriculture play an important role in decision making at both micro and macro levels. Generally, higher prices are expected to result in a larger output, which indicates that the farmers make appropriate adjustments in acreages under crops in response to changes in the prices.

According to Raj Krishna[1], agriculture is responsive to price changes. Farmers respond favourably to price changes. Prices are the most important determinants of the area under different crops. Therefore, agricultural price policy should be positive. A positive price policy as a part of growth policy has three important functions. They are:

(1) To accelerate or decelerate the growth of output of individual crops or in the context.

(2) To accelerate the growth of agricultural output as a whole,of planning to steer the crop-mix according to targets, and

(3) To secure adequate increase in the marketed supply of food crops in countries where a large part of output is retained by the peasants for home consumption.

The objectives of agricultural price policy in developed and developing countries are not similar. In the developed countries like U.S.A., the agricultural price policy aims at to bring the agricultural income to the level of non-agricultural income. According to Raj Krishna[2] the main aim of the agricultural price policy in North America and West European countries is to support farm incomes rather than to provide incentives for increase in agricultural production. According to Bhise[3], in India the agricultural price policy aims at to stabilize the levels of agricultural prices in order to protect the interests of the producers from a sharp fall in agricultural prices and that of the consumers from an excessive price rise. Agricultural price policy in India is almost formulated to achieve the desired output levels in different crops. Agricultural growth which is the fundamental basis of overall economic growth requires sound agricultural policy. Agricultural price policy, therefore, is one of the important instruments in the hands of the government to achieve higher growth rate in agricultural production. It also helps for reshaping allocation of resources and distribution of income both within the agricultural sector and between agricultural and non-agricultural sectors. This is more applicable to a country like India which is suffering from chronic shortage of foodgrains.

According to Satish Chandra[4], in a viable economy like India, agricultural price policy plays an important role for eradication of poverty and quickening growth so that the standard of living of the poor people of rural areas, which is of great significance, can be improved.

Rajbans Kaur[5] quotes that the developed economies have low income elasticity of demand for agricultural products in general and foodgrains in particular and have high price elasticity of supply of agricultural products. In contrast, agricultural economies in the developing countries (like India) are characterised by relatively inelastic aggregate demand and supply elasticities which cause sharp fluctuations in the agricultural prices. So it is necessary to moderate the price fluctuations and induce investments in agricultural sector which stimulate productivity and development of the country. So agricultural price policy has become a common element in economic management of both the developed and developing countries now-a-days.

In order to make price policy an effective instrument for inducing desired changes in the resource allocation to different crop enterprises, the knowledge of farmers supply responsiveness is essential. If price of a crop is high then there is a tendency to increase the acreage under that particular crop. So, the response of the production of a particular crop is influenced by the possibilities of engaging in other lines of production. If the opportunities in alternative lines of production are limited, it is likely that the producers of the particular commodity will respond less quickly to given changes in price than they will if numerous alternatives are available. Therefore, it wants to restructure the relative price in such a way that the output of the main crop increases at positive rate but at the same time the targeted level of output of alternative crops is not affected. As the prices are the barometer of the health of the economy (they indicate the strength and weakness of the economy), the planners must know the behavioural supply response functions for both the main crops as well as the alternative crops. Price is most important determinant of profit or loss in the farm enterprise[6].

Jhala[7] said that the farmer is a central point who makes the final decision regarding the allocation of land to different crops depending on his expected revenue from different crops, assuming input costs are either the same or move uniformly over time for different crops, the expected revenue depends on expected prices and expected yields. If yield levels are constant over time due to lack of any significant technological change the acreage response equals the output response.

According to Nerlove Marc[8], the important decisions relating to which crop and how much crop to grow on how much land with what levels of non-land factors of production etc. are essentially based on the relative profitability of each crop. Any change in the levels of production of different crops is thus bought about through inter-crop shifts on land, owing to varying probability from crop to crop. Bhagat[9] stated that the price may be the only consideration, if the yield rate remains unchanged over the period. If the yield rate varies, price may not be the sole decisive factor. In fact, when the technological development takes place, the year to year changes in the land allocation reflect changes in the farmers' decisions not only owing to variation in prices but also because of variations in the yield rates. Again if the farmers produce the crop mainly for the market, price is an important consideration for them but the subsistence farmers may attach more importance to the yield rates because most of the crops are grown for their own consumption. The consumer-producer farmers might also consider the price and yield rate both simultaneously and are interested in income from the crop, specially the net cash income.

Narayanan and Parikh[10] also emphasized in their study that the producer-consumers give due importance to relative profitability i.e. relative gross return while allocating land to specific crops. Even if the crop is grown mainly for self-consumption, the farmer should not be insensitive to relative profitability. A study of the year to year variations in acreage of different crops indicates that farmers continuously make shifts in their production programme. Variations in acreage depend on various factors. Price of the commodity is one of the important factors as well as the prices of alternative crops, damage from disease and insect pests, changes in the cost of production, improvements in the varieties, cost of competing crops and weather conditions which cause relatively high or low yields.

According to Ravikumar, Sreelakshmi and Raju[11], prices determine not only what shall be produced, but also how much should be produced and their distribution for achieving the targets of agricultural production. It is important that the farmer should have full confidence that additional efforts and investments will yield adequate return. The wide fluctuations in the prices affect the farmers capacity of making sustained efforts for increasing production. These fluctuations in the prices of agricultural produce are the greatest hurdle in the way of agricultural development. It is therefore, necessary to conduct the price-response studies which form an essential requirement for policy formation. On the basis of price response studies, the farmer can make decisions for proper allocation of a crop area by anticipating price in the future. Kamala Devi and Rajgopalan[12] explained that the response of prices on acreage cannot be judged by a study of the prices of that particular crop alone. It should be studied in relation to the prices of other crops which might have an influence on the particular crop. Where competing crops are involved the equilibrium price tends to vary with the prices of competing crops.

Generally it is hypothesized that an increase in price induces farmers to increase the area under cultivation to the best possible extent, to apply more inputs to produce the maximum possible quantum using the best available technology. Economists and policy makers are interested in the relationship between price changes and the supply response of primary products in the developing countries as the farmers of such countries are responsive to economic incentives such as prices of agricultural commodities and agricultural inputs are main features of the underdeveloped countries[13]. It is essential to develop policies in order to avoid the deficits in the balance of payments, unemployment, deterioration in the commodity terms of trade and reduction in the earning of agricultural exports, in framing policies relating to government price support programmes, exchange rate, tariffs and producers association, the knowledge of supply response to price changes is must. The question of farmers response to price signals has been extensively examined by a number of researchers in recent years.

However, the researchers generally used two types of models in estimating acreage response of the farmers. The first model is called as 'Traditional price lag model' and the second model is 'Adjustment lag model', which is also known as Nerlovian model[14]. Traditional price lag model assumes that the farmers instantaneously and fully adjust their acreage allocation in response to the changes in lagged prices. Nerlovian model which postulates that the actual acreage under a crop in any period is adjusted in proportion to the difference between the desired acreage in the long-run equilibrium and the actual acreage in the preceding year. In fact, the original formulation of the Nerlovian adjustment model hypothesizes acreage as a function of expected prices but the expected price cannot be observed in practice. However, the expected prices in any year can be expressed as a function of actual price in the last year and expected prices in the last year. We can replace last year's expected price by a linear function of last year's acreage. Thus through these algebraic substitutions, the final form of adjustment model expresses acreage in any year is a function of last year's actual price and last year's acreage and thereby ignores the effect of expectational lags in prices, if any.

Among the researchers[15] those who have used the Nerlove's supply model, there has been a wide diversity of interest which has occasioned considerable adoption in the form of the basic model, depending on the crop, the period and the country being considered. However, to measure the effects on cultivator marked responsiveness of a number of economic, social and political factors, estimation of supply elasticities of several crops were made by using a Nerlovian type model. This approach has been employed on numerous earlier occasions by the authors and researchers interested in Indian agriculture.

Although the farmers response to price changes has been the subject of several investigations, the nature and extent to which farmers respond to such changes remains a widely debated and controversial issue. All the same, the fact cannot be denied is that the price changes do influence farmers decision making regarding allocation of resources[16].

Thus, the farmers responsiveness to the price changes is helpful in formulating an appropriate price policy which is of immense importance for the growth and stability of the production on one hand and economy on the other.

Commercial crops play an important role in the development of a country like India. India is one of the most leading oilseeds producing country of the world. Oilseeds play a dominant role in the economic development of the country and it is an important base for the agro-industries such as oil milling industry, soap industry, hair oil industry, paint industry, toiletries industry etc.

Oilseeds are one of the age old products of this earth and they play an important role in the human diet. Oilseeds provide oils and fats which impact nourishment and warmth to human system. Oilseeds are multi-product and multipurpose commodity and their contribution is multifarious in our national economy.

With the introduction of high yielding varieties of wheat after mid-sixties and of rice in early seventies, these two crops became markedly more profitable to grow compared with other crops in the areas where assured irrigation facilities were available. This became instrumental in diverting some area also from other crops (including oilseeds) to these two crops. Also, the use of water and fertilizers got diverged from competing crops to these two crops. This continuous change spread over two decades solved the food problem of the country in respect of cereals, but, inter alia, diversion of irrigated crop lands at the margin from under oilseeds did not permit the growth of oilseeds production to match the domestic demand for edible oils. In India[17], the total area under oil seeds production (nine major oilseed crops) has increased to 23,250 thousand hectares in 2000-01 from 10,727 thousand hectares in 1950-51 i.e. an increase of 2.2 times; whereas, the production and yield increased to 18,400 thousand tonnes and 791 kg per hectare in 2000-01 from 5131 thousand tonnes and 478 kg per hectare in 1950-51 i.e. registering an increase of 3.6 times and 1.7 times respectively.

Although India is one of the largest producer of oilseeds, the edible oils production falls short of the country's requirement. This has necessitated large scale imports of edible oils.

About 10 per cent of the total cropped area in the country estimated to be under oilseed crops, the more important of which are groundnut, rapeseed, mustard, sesame, linseed, castor and soyabean. About 92 per cent of the total land under oilseed crops is rainfed and hence the crop prospects are uncertain.

Economists have extensively investigated the growth performance of rice and wheat during the last four decades. It has been widely researched at the global, regional, national, state and household levels. Unfortunately, scant attention has been paid to the study of oilseed crops, which play an important role in sustaining crop systems and meeting the domestic needs of edible oil in the country. Therefore, an attempt must be made to provide current evidence on the temporal and spatial dimensions of the oilseed development. The production of oilseeds like other agricultural commodities is primarily determined by acreage. Therefore, it would be worthwhile to investigate the factors influencing acreage. Various price and non-price factors influence the farmer's decisions regarding land allocation to different crops. The first segment includes input and output prices. These range from last year's harvest price of the crop, availability of minimum support price, last year's harvest price of the competing crop to prices of fertilizer, power, seed, water, insecticides and availability of credit. Similarly, a host of non-price factors also play an important role. The major factors are last year's acreage and yield, availability of improved seeds and irrigation, rainfall, facility of procurement by government agencies, resistance to crops to pest attacks, extension services, home consumption, availability of alternate crops, credit and assured market. Unfortunately, crop specific information on all these variables is lacking and therefore, even tabular analysis is not possible. Owing to this difficulty, the findings of acreage response studies are based on a few variables for which data are available. Thus, factors affecting oilseed acreage are numerous and their contribution is generally estimated through an acreage response model. Most of the studies cover a period up to eighties but policy scenario has dramatically changed after gradual liberalization of agriculture. Moreover, the nature of acreage response differs from crop to crop and region to region.

Marathwada is the most backward region of Maharashtra state and agriculture is the most important sector of the region's economy. It is endowed with natural resources. The natural factors such as rainfall and soils are relatively more favourable in this region. The soils and climate of this region are favourable to take oilseed crops such as groundnut, sunflower, safflower and recently soyabean. Marathwada shares 12.5 per cent of the states total area under oilseeds and around 11 per cent of state's total oilseed production. There is more scope in this region to increase area under oilseeds, especially groundnut, sunflower and safflower. The acreage allocation decisions of farmers are always made by comparing relative returns. Further, rainfall, irrigation and some other factors also cause the acreage allocation decisions of the farmers. It will, therefore, be interesting to probe into the factors which determine the acreage under major oilseeds in Marathwada. No such study has so far been conducted for major oilseed crops in Marathwada for the eighties and nineties. This study also aims at to study the growth in oilseeds production and factors which are important in making decisions regarding allocation of area to oilseed crops in Marathwada. More precisely, this study aims at to measure the growth and acreage response of oilseed crops in Marathwada region during the period from 1980-81 to 1999-2000.

The following are the main objectives of the present study:

1. To study the growth in area, yield and production of oilseed crops in Marathwada.
2. To study the area response of important oilseed crops in Marathwada.
3. To probe into the factors which are important in determining acreage of oilseed crops in Marathwada.
4. To suggest policy measures to increase acreage and production of oilseed crops in the Marathwada region.

RESEARCH METHODOLOGY

The present study pertains to oilseed crops grown in the Marathwada region of Maharashtra state. This study covers the period from 1980-1981 to 1999-2000. It covers four important oilseeds, namely groundnut, safflower, sunflower and soyabean. The required time series data pertaining to area, production and yield of oilseed crops, farm harvest prices, rainfall, irrigation and prices of its competing crops have been collected from the Season and Crop Reports of the Maharashtra State, Districtwise General Statistical Information of Agriculture Department (Epitome of Agriculture Part I and Part II), Socio-Economic Review and District Statistical, Abstracts of different districts of Maharashtra state.

THE MODEL

In order to study the growth in area, production and yield of oilseed crops, the annual compound growth rates are calculated by using exponential function as follows:

$$Y = AB^X$$

$$r = (B-1)\ ´\ 100 \text{ (in percentage)}$$

where,

$$Y = \text{Area/yield/production}$$

$$A = \text{constant}$$

$$B = 1+r$$

r = compound growth rate (CGR)

The area response model developed by Nerlove is used to study the area response of oilseed crops in Marathwada region. The model (modified by others) used is as under:

$$A_t = a + b_1 A_{t-1} + b_2 RP_{t-1} + b_3 RY_{t-1} + b_4 I_{t-1} + b_5 RF_t + U_t$$

where,

A_t = Area in hectares under the crop in the year t,

A_{t-1} = Area in hectares under the crop in the year t-1,

RP_{t-1} = Relative price in the year t-1,

RY_{t-1} = Relative yield in the year t-1,

RF_t = Sowing period rainfall (mm) in the year t,

I_{t-1} = Gross irrigated area in the year t-1.

The independent variables used in the area response equation are lagged acreage, price, yield, irrigation and rainfall. Different forms of regression equations are employed on the data (semi-log, double log and linear).

SCHEME OF CHAPTERS

The present book consists of six chapters. A brief introduction of each chapter has been given as follows:

The first chapter covers the importance of the present study, the importance of supply response studies, the objectives and methodology employed.

Review of literature on acreage response of oilseed crops has been taken in the second chapter.

The third chapter explains agricultural economy of Marathwada region.

The fourth chapter is devoted to discuss the growth performance of oilseed crops in Marathwada region in terms of area, yield and production.

The fifth chapter deals with the study of acreage response of oilseed crops viz. groundnut, sunflower, safflower and soyabean in Marathwada region.

The sixth chapter presents the findings of the present study, and suggests appropriate policy measures.

NOTES AND REFERENCES

1. Krishna, R. A. (1967), Price Policy and Economic Development in H.M. Southworth and B.F. Johnston (edt.) Agricultural Development and Economic Growth, Cornell University Press, Ithaca, New York.
2. *Ibid.*
3. Bhise, V. B. (1988), Price Policy for Foodgrains in India in G.S. Kalyankar edited Indian Economy, Marathwada University, Aurangabad.
4. Satish Chandra (1985), Agricultural Price Policy in India, Chug Publication, Allahabad.
5. Kaur Rajbans (1975), Agricultural Price Policy in Economic Development, Kalyani Publishers, Delhi.

6. Krishna Raj (1963), Farm Supply Response in India–Pakistan: A Case Study of Punjab Region, *The Economic Journal*, Vol. LXXIII, No. 291, September.

7. Jhala, M. L. (1979), Farmers Response to Economic Incentives: An Analysis of Inter-Regional Groundnut Supply Response in India, *Indian Journal of Agricultural Economics*, Vol. XXXIV, No. 1, January-March.

8. Nerlove Marc (1958), The Dynamics of Supply: Estimation of Farmers Response to Price, John Hopkin Press, Baltimore.

9. Bhagat, L. N. (1985), Determinants of Wheat–Acreage Fluctuations in Bihar: A Study of Temporal-Spatial Variations, *Indian Journal of Agricultural Economics*, Vol. XL, No. 2, April-June.

10. Narayana, N. S. S. and Parikh, K. S. (1979), Agricultural Planning and Policy in Draft Sixth Plan: Will Farmers Fulfill Planners Expectation? *Economic and Political Weekly*, Vol. XIV, No. 31, 31 and 32, Special Number, August.

11. Ravikumar, K. N., Sreelakshmi, K. and Raju, V. T. (2000), Price Response Analysis of Agricultural Commodities in the Selected Regulated Markets of Andhra Pradesh, *Agricultural Situation in India*, Vol. LVI, No. 12, March.

12. Kamala Devi, P. and Rajgopalan, R. (1965), Price and Acreage Response (A Case Study of Groundnut Crop in North Arcot District), *Indian Journal of Agricultural Economics*, Vol. XXI, No. 1, January-March.

13. Pal, S. P. (1975), Supply Response and Optimal Agricultural Policy in a Planned Underdeveloped Economy, "*Indian Journal of Agricultural Economics*, Vol. XXX, No. 2, April-June.

14. Nerlove Marc (1958), Distributed Lags and Estimation of Long Run Supply and Demand Elasticities: Theoretical Consideration, *Journal of Farm Economics*, Vol. XL, No. 2, February.

15. Nerlove Marc: *Ibid.*

16. Krishna Raj (1963), Rapporteur's Report on Economics of the Cropping Pattern, *Indian Journal of Agricultural Economics*, Vol. XVIII, No. 1, January-March.

17. Padmavathi, A. (2006), Trend Analysis of Oilseeds in India, *Southern Economist*, Vol. 45, No. 13, November 1.

2 Review of Literature

In this chapter, an attempt has been made to review literature in regard to acreage response of oilseed crops. Number of investigations pertaining to this problem have been brought out so far. Our task here, therefore, is to take review of some of the important acreage response studies, particularly in Indian agriculture. The review covers main studies on crop acreage response which have used Nerlovian adjustment lag model and the oilseed crops which are considered in the present study.

Padwal[1] observed in his study of production and price trends of major oilseeds in India that, when acreage increased, the production has not increased in the same proportion resulting in abnormal fluctuation in prices.

Nerlove Marc[2] explained the behaviour of the farmers in USA. He estimated the supply responsiveness of individual crops namely cotton, wheat and corn crops and found the coefficient of adjustment to be 0.51 for cotton, 0.52 for wheat and 0.54 for corn crops. Thus, he came to a conclusion that the farmers are responsive to changes in relative prices during the period 1909 to 1932 in USA.

Misra and Sinha[3] have pointed out that in a country like India due to rigid factors of agricultural production, total agricultural production is not expected to respond quickly to price incentives. In case of commercial crops such as jute, cotton and sugarcane higher relative prices may increase acreage of crops with lower relative prices. But the same is not true in case of foodgrains production or total agricultural production.

The scope for an increase in the total cultivated area due to higher prices is limited in India. As most of the farmers are subsistence farmers, they cultivate mainly foodgrains for family consumption and other family expenses. The foodgrains area does not increase or decrease very greatly with the rise and fall in prices of foodgrains.

Raj Krishna[4] has examined the supply response to price for some agricultural commodities in undivided Punjab. The study indicates that price alone is important factor in determining the acreage under maize and sugarcane. In case of cotton price is more important factor than yield (in terms of elasticities). In the case of cotton, bajra and wheat (irrigated) irrigation capacity is more important than price. Yield is more important than price in the case of rice and rainfall is the most important in case of unirrigated wheat, barley and gram are the major Punjab crops, jowar is the only crop which has shown a negative response to relative price movements.

John[5] has examined three different types of responsiveness i.e. (1) changes in the area and output of a commodity in response to changes in the price of that commodity, (2) changes in the area and output of a commodity in response to changes in the prices of that commodity relatively to that of another single commodity grown as its substitute, and (3) changes in the area and output of a commodity in response to change in the price of that commodity relative to change in the average price of a group of commodities grown as its substitutes. The results of the study indicate that in the majority of cases positive relationship of one kind or another exist between price and area/output or relative price and relative area/output.

Kamala Devi and Rajgopalan[6] examined the influence of relative prices of groundnut and its competing crops on acreage in North Arcot district, Madras state, as a major groundnut growing district. The period studied was 1934-35 to 1961-62. The acreage figures for the period 1934-35 to 1953-54 were collected from the quinquennial statistical abstract for the Madras province, and for the latter years from the season and crop reports of Madras state. Wholesale prices were also collected from some sources and for same period. Firstly, an attempt was made to find out the relationship between the acreage under groundnut and the acreage under competing crops. For this purpose the groundnut acreage was examined in relation to the total acreage of all the other competing crops viz., ragi, varagu, samai and pulses. Year to year fluctuations in the acreage of these two had been worked out and it was seen that there is an inverse relationship between the two. With only a few exceptions in all the other years an increase in groundnut acreage had caused a decrease in the acreage under competing crops and vice versa.

The degree of this relationship had been explained by statistical computation using linear equation $y = 88.1 - 0.80\,X$

Where,

X = Change in total area of ragi, varagu, samai and pulses in hundreds of acres and

Y = Change in area of groundnut in hundreds of acres.

The result is that a good representation of the relationship between the change in the total area of the competing crops and change in the area of groundnut. An increase of 100 acres under the competing crops showed a decrease of 80 acres under groundnut.

The response of prices on acreage cannot be judged by a study of the prices of that particular crop alone. It should be studied in relation to the prices of other crops which might have an influence on the particular crop, where competing crops are involved the equilibrium price tends to vary with the prices of competing crops.

The area of groundnut is related to the relative price of groundnut to ragi in the preceding year.

The linear regression showed the following results:

$$Y = 82.10 + 61.12\,X,$$

$$\text{Where, } X = \frac{\text{Groundnut price}}{\text{Ragi price}}$$

Y = Area of groundnut of next year in thousands of acres.

The effect of relative prices on yield per acre has been examined. An increase in price, both absolute and relative, might induce the farmer to increase the productivity of that crop by increasing the inputs, even without increase in acreage. It is examined that a rise in the relative prices does not cause a rise in productivity. Its influence was only on the acreage planted. This analysis suggests that the influence of relative prices on productivity is mainly through shifts in acreage between competing crops and not through changes in production.

In short,

(i) increase or decrease of acreage under groundnut in North Arcot is inversely associated with the acreage under its competing crops. An increase in the area under the competing crops would mean a decline in the area under groundnut.

(ii) An increase in the relative price influences the acreage under groundnut in the following year. The ratio of groundnut price to ragi price is found to have an influence in the acreage under groundnut in the following year.

(iii) The influence of relative prices on productivity is not at all significant. A rise in prices does not cause an increase in productivity. It is only a change of inputs from one crop to the other.

Dharam Narain's[7] study, though not based on rigorous econometric analysis, was an important and extensive study of supply response of several major crops grown in different regions of India. It covers the period from 1900 to 1939. His analysis indicates the existence of a stronger positive relationship between changes in acreages and in prices in case of non-food crops than that in food crops.

Dharam Narain in his study relating to the impact of price movements on areas under selected crops in India has found that the Indian farmer is significantly responsive to price. He finds a striking difference of behaviour between cash crops and food crops. The study indicates that the Indian cotton grower is about as significantly responsive to commercial stimulus as his counterpart, engaged in highly commercialized agriculture in the United States of America. It has been observed that in the case of foodgrain crops, rainfall assumes that status which price does in the case of cash crops. The study also demonstrates the role of price in foodgrain areas.

Rajagopalan[8] examined three variants of the basic Nerlovian supply model for Madras to estimate supply parameters for groundnut output between 1939 to 1961. He found stronger and more consistent evidence for a significant role of prices in influencing groundnut output than it did for any of the other crops he considered. Short-run elasticity estimates were 0.68 and 0.74 for the Northeast and interior regions, respectively. While cross elasticities with respect to price in the two regions were –1.06 to –0.99. For groundnut, he used two variants of the basic Nerlovian model to estimate supply acreage expressed in terms of (a) lagged groundnut prices deflected by an index of substitute crops (ragi and rice) prices, lagged (one and two periods) acreage and a trend variable, and (b) lagged absolute groundnut prices and substitute crop prices entered separately, lagged per acre groundnut yield, lagged acreage and a trend variable.

The results showed that though little statistical significance was found for the role of groundnut prices in output, except in the southeast region where there was a much less important crop, the previous price pattern was clear in the major districts and was further accentuated by the positive parameter estimated for substitute crops in the north-east region and the negative yield parameters in all the three regions. The groundnut price elasticity estimates were -0.50 in the north east region and a cross elasticity of –0.87 in the interior region.

Behrman[9] in his study relating to supply response in underdeveloped agriculture, a case study of four major annual crops in Thailand has found that farmers in underdeveloped countries respond rationally and substantially to economic incentives. He has also found that there is no significant evidence for the hypothesis that institutional constraints preclude significant response to economic incentives in underdeveloped countries. The study indicated that large variations in prices had an adverse impact on acreage.

Reddy[10] observed responsiveness of farmers to the changing circumstances, namely change in relative price level, relative yield and rainfall in allocating their acreage among groundnut and competing crops like cotton and jowar in Karnool district of Madhya Pradesh. This study was related to the period 1931-43. The study concluded that the farmers in the Karnool district under study were responsive to price changes and also to relative yield changes. The Nerlovian coefficient of adjustment indicated that farmers were readily adjusting to changing circumstances. The negative coefficient of trend indicated a downward shift in allocation of land to groundnut.

Kaul and Sidhu[11] attempted to examine the acreage response to prices for major crops viz., wheat, paddy, maize, groundnut and desi cotton in Punjab state. Their study pertains to the period 1960-61 to 1969-70. The study was an attempt to obtain the best estimates of the response of Punjab farmers to prices while making a decision about acreage allocation to crops. They used Nerlovian type of model for this study. The main postulation of this study was that for decision-making in regard to the allocation of acreage, farmers look to harvest prices or any other price. As such harvest prices exercise exceedingly more influence in the decision of the farmers than wholesale price. The farmers base their expectations of future prices on harvest price. The second postulation of this analysis was that the decision of allocating the acreage does not depend upon past prices alone but the extent of variation in these prices count much in building these price expectations to the farmer. The index of value productivity (yield x price) of the crop has been deflated by the value productivity index of the competing crops and this has been used as a variable instead of price. The competing crops for the crops were as follows: Gram and barley for wheat; maize for paddy; bajra, cotton (A), cotton (d) and sugarcane for maize; cotton (A), sugarcane, maize and wheat for cotton (d) and bajra for groundnut.In connection to wheat the coefficient of lagged acreage was significant and the value of R^2 was 0.9144 to 0.9203 and long run elasticity was higher than the short run elasticity. In the case of groundnut the coefficient attached to the lagged acreage was significant. There was not much of difference in so far as the value of R^2 was concerned. The long run elasticities with respect to relative profitability worked out to be highly elastic for groundnut. It ranged from 3.057 to 3.251. The reason of this high elasticity was that it was a cash crop and marketable surplus in cash crop was higher than that in foodgrain crops .The coefficient of lagged acreage was significant in connection to cotton (d). The value of R^2 was 0.8756 to 0.9632. The long run elasticity with respect to relative profitability was higher than short run elasticity and was elastic because cotton (d) was a cash crop and the marketable surplus was very high.

From the above results it is evident that the coefficient of variation as a variable in place of standard deviation has consistently given higher coefficient of determination. The long run elasticities were higher than the short run elasticities in cash crops viz. groundnut and cotton (d) because these are high risk crops whereas paddy and wheat were relatively low risk crops.

Mishra and Radhakrishnan[12] examined the price effect on acreage response for groundnut and bajra in Gujarat state. Their study was related to the period 1949-50 to 1968-69. They used four Nerlovian type adjustment lag models, two traditional models and two Nerlovian type models, allowing scope for directional differences in the long run price effects. The explanatory power of the model was found to be satisfactory and the signs of coefficients of almost all explanatory variables were the expected ones.

Sahay[13] carried out a study of acreage response of groundnut in major (9 zones) groundnut zones for the period 1954-55 to 1967-68. He applied Nerlovian adjustment lag model for acreage response of groundnut. The important finding of the study was that the best price expectation for acreage response of groundnut in different zones was not the same. A high value of coefficient of adjustment indicate lesser rigidity in adjustment of output. Generally, low value of this coefficient was found and it took 6-8 years for 95 per cent of the influence of price change to be realized.

The findings established the relationship of farmers' response to changes in relative prices but the extent of response was found to vary between the zones.

Parikh's[14] study included analysis of some of the major Indian groundnut regions. He used several variants of supply model, the results from testing with groundnuts data were reported. Average expressed as a function of (a) lagged groundnut prices deflated by a general price ratio and lagged area, (b) lagged deflated prices, yield acreage and rainfall at sowing period, (c) lagged deflated prices, lagged deflated proceeds per acre, and lagged (one and two periods) area, and (d) lagged (one and two periods) relative price of groundnuts to competitive crop prices, weather and acreage.The results of Parikh's regression analysis were that positive responsiveness indicated in all three provinces, though elasticity values are not quoted for Madhya Pradesh, where the Boon-Raung study indicated negative response during a more recent time period. For Bombay, Parikh's estimates are somewhat high even for short-run elasticity when compared with other results for the period 1914 to 1939.

Madhavan's[15] study pertains to the post-war period (1947-1965) for which he considered all regions of Tamil Nadu and used Nerlovian log model expressed in logarithmic terms with crop acreage as a function of lagged prices relative to those of a competing crops, lagged yield and acreage of the oilseed and the competing crops and an index of rainfall in sowing period.

In his multicrop study of Tamil Nadu considered the supply of four food crops viz., rice, Sorghum, ragi and cumbu and four cash crops viz., cotton, sugarcane, groundnut and sesamum. He included in each regression only a single competing crop, not of as a weighted average of all so designed alternatives. He used the model of the following type.

$$\log X_{it} = b_o + b_1 \log \frac{P_{it-1}}{P_{jt-1}} + b_2 \log E_{it-1} + b_3 \log E_{jt-1} + b_4 \log$$

$$X_{jt-1} + b_5 \log X_{it-1} + V_{it}$$

Where,

$\frac{P_{it-1}}{P_{jt-1}}$ = Relative Price of crop i with respect to crop j, lagged one year,

E_{it-1} = Yield per acre of crop i, lagged one year,

E_{jt-1} = Yield per acre of crop j, lagged one year,

X_{it-1} = Acreage of crop i, lagged one year,

X_{jt-1} = Acreage of competing crop, lagged one year.

Madhavan considered cumbu, Ragi, Sorghum, cotton, Sesasum and sugarcane to be competitors of groundnut and cotton and groundnut to be competitors of Sesasum. He also found generally more statistical, significance for the price coefficients of oilseed than he did for the other food and cash crops he considered. The role of prices in sesamum output seems particularly notable.

Madhavan's study reveals that the Tamil Nadu farmers in general seem to respond to relative prices, but the degree of response varies from crop to crop. In general the crops which are grown for sale such as sugarcane, sesamum, groundnut and cotton seem to respond more to relative prices than subsistence oriented crops such as rice, sorghum, ragi and cumbu. Acreage elasticity estimates with respect to price are found to be generally higher when both dependent and competing crops are from the commercial crop group and low when both are found the cereal crop group. Elasticities associated with a dependent commercial crop and competing cereal crop are found to be generally higher than those associated with a dependent cereal crop and a competing commercial crop.In almost all the cereal crops and even in some commercial crop acreage elasticity with respect to yield is higher than acreage elasticity with respect to price.

Mahapatra's[16] study pertains to the period 1956-57 to 1968-69 in Orissa state. He used Nerlovian adjustment model to study the impact of price changes on acreages of selected crops viz., rice, jute and sugarcane. His study concentrating on the two districts of Orissa state, Mohapatra found that price does not play an important role in determining the acreage under the crops due to the low land-man ratio, small holdings and the use of traditional methods.

The study of Gupta[17] indicates that the price is not a major factor in explaining the variations in the acreage under various crops. His study does not support the positive price response hypothesis as well as negative supply response hypothesis in case of Madhya Pradesh state. He found that the rate of increase in area under crop is higher in case of those crops which have less variability in prices over time. Productivity per acre has significant influence in determining the acreage under rice and groundnut. Trend has significant influence in determining the acreage under wheat, gram and groundnut. It explains that technological changes, institutional changes, extension of irrigation and other factors assume an important place in the agricultural growth.

A study by Singh, Parmatma Singh and Raj[18] is related to the period 1950 to 1970. This study is related to the cultivation of cereals–wheat, barley, maize, gram and bajra as well as cash crops namely cotton (d) and rape and mustard seed. They used simple explanatory model and represented area in terms of harvest price, rainfall in the pre-sowing period and agricultural technology. The latter was expressed as yield per acre, rather than the simple trend variable used in other studies of India.

They reported fairly sizeable positive supply elasticities for four crops that were derived from statistically significant estimated price coefficients. Rainfall coefficients were positive and significant in four crops i.e. wheat, barley, rice and bajra. But in only a single case (maize) did the yield proxy for technology, show significance. In case of cotton, only the price coefficient proved significant in the regression estimate and value of short run elasticity of cotton is 0.622.

Askari and Cummings[19] used the Nerlovian model of the following three equation form to discuss some of the limitations of the Nerlovian model.

$$(1)\ A_t^D = \alpha_0 + \alpha_1 P_t^E + \alpha_2 Z_t + V_t$$

$$(2)\ A_t = A_{t-1} = \gamma \left(A_t^D - A_{t-1}\right)$$

$$(3)\ P_t^E - P_{t-1}^E = \beta\left(P_{t-1} - P_{t-1}^E\right)$$

where, A_t and A_t^D are actual desired acreage planted in period t,

P_t and P_t^E are actual and expected prices in period t and Z_t represents other than non-price factor(s) hypothesized as significantly affecting supply.

Desired output is thus expressed as a function of expected prices; though both are non-observable variables, equation (2) and (3) define each in terms of past observed values. Since inflexibilities may hinder the cultivator from making in one period exactly the acreage changes he may desire, the actual change, in any given period, is expressed as a portion of the desired change.

Price expectations are also framed in adaptive terms with expectations shifting in each period by some fraction of the difference between the recent actual and expected prices. A number of non-price variables has been included in the basic supply equation, most commonly rainfall and crop yield indices, and simple trend variables indicated to represent the effects of monotonic changes in both technology and/or in the supporting infrastructure. The author, in order to avoid the problem of efficiency and consistency in estimation used to maximize the likelihood function of the observation with respect to the b's. Under the assumption that the disturbance terms are distributed N(O,σ²I), they followed the log of the likelihood function:

$$L\left(\underline{A}\,\underline{X}, b, \sigma^2\right) + \tfrac{1}{2}\log(2\pi) - \tfrac{1}{2}\log\sigma^2 - \tfrac{1}{2}\sigma^2(\underline{A} - \underline{XB})'(\underline{A} - \underline{XB})$$

The likelihood function is maximized when the sum of the squared residuals is minimized. And they obtained estimation of the b's by solving:

$$\partial W'W / \partial b_i = 0$$

These estimates are consistent as sympathetically unbiased and efficient.

Askari and Cummings lastly concluded that the value of the Nerlovian model as an analytical tool must certainly be recognized. Recognition however, must also be made of the limitations on what can be gained from this kind of supply analysis and concluded that the anticipated yield has great bearing on the farmers decision making. Thus they advocated a policy for increasing agricultural acreage and output beyond price incentives.

Acharya and Agrawal's[20] study pertains to Rajasthan for the period 1959-60 to 1972-73. They used Nerlovian model in their study and considered price, weather, rainfall, HYVs as the independent variables for food crops and commercial crops. They found that the relative prices have significant effects on the area under oilseed crop viz. groundnut.

Acharya and Bhatiya[21] in their study used regression analysis to estimate acreage response to price, yield and rainfall for wheat, gram, maize, bajra, groundnut, sugarcane and cotton crops in Rajasthan. This study was related to the period 1952-53 to 1970-71. Modified Nerlovian model with price variability, yield and rainfall as the independent variables were considered to be more appropriate. All the crops had positive short-run and long-run acreage price elasticities. The study also revealed that farmers in Rajasthan considered 2 to 7 years past price in acreage allocation decision.

Sidhu and Prakash Mehta[22] studied the response of farmers to prices in a case study of potato and groundnut in Punjab for the period 1950-51 to 1970-71 and concluded that farmers allocated the acreage under potato by observing the harvest price of wheat crop. In case of groundnut, relative returns turned out to be significant. The farmers take into consideration the returns which they expect from competing crop. The absolute prices of competing crops did not influence the allocation of hectarage of groundnut.

Jhala[23] used Nerlovian adjustment lag model. The Nerlovian type model depicting the farmers' behaviour in groundnut supply response for a more or less uniform in the period 1950-51 to 1970-71 employing a common estimating procedure and a common approach to the specification of variables, especially that of price variable. Price, yield, area and rainfall were the independent variables.

Simplest form of Nerlovian lag model was as follows:

$$A_t^{\bullet} = b_0 + b_1 P_{t-1} + b_2 Z_{t-1} + b_3 R_t + U_t \quad (1)$$

$$A_t - A_{t-1} = B\left(A_t^{\bullet} - A_{t-1}\right) 0 < B < 1 \quad (2)$$

$$A_t = C_0 + C_1 P_{t-1} + C_2 Z_{t-1} + C_3 R_t + C_4 A_{t-1} + V_t \quad (3)$$

The first equation is a behavioural relationship stating that the desired groundnut acreage ($A_t^{\bullet}$) depends on relative farm prices of groundnut lagged one year (P_{t-1}), relative yields of groundnut lagged one year (Z_{t-1}) and sowing period rainfall (R_t). Equation (2) is a partial area adjustment equation in which B is coefficient of adjustment and equation (3) is the reduced form of above two equations, which eliminates the unobserved variable ($A_t^{\bullet}$) by an observed variable (A_t). The coefficients and error terms of the equation (3) are related to those of equation (1) and to the coefficient of adjustment as follows:

$C_o = b_o B$, $C_1 = b_1 B$, $C_2 = b_2 B$, $C_3 = b_3 B$, $C_4 = (1\text{-}B)$, $V_t = Bu_t$

The variables used in the estimated equation are presented below. Instead of relative yield variable, yield of groundnut and competing crops have been introduced separately as the weighed average yield of cotton and bajra might not make any sense.

A_t = Acreage under groundnut in the concerned region (in '000' acres)

P_{t-1} = The farm harvest price of groundnut deflated by the weighed average farm harvest price of competing crops in the concerned region.

Z_{t-1} = Average groundnut yield in the region (in kg/ha),

Z_{t-1}^{J} = Average yield of jowar (in kg/ha),

Z_{t-1}^{B} = Average yield of bajra (in kg/ha.)

Z_{t-1}^{C} = Average yield of cotton (in kg/ha.)

Z_{t-1}^{M} = Average yield of maize (in kg/ha.)

Z_{t-1}^{R} = Average yield of ragi (in kg/ha.)

R_t = Rainfall in sowing period in current year (in mms).

Jhala selected 81 districts out of 111 Indian groundnut producing districts forming into 14 regions. Thus the scope is much wider. He used single uniform price relative i.e. farm harvest price of a crop deflated by weighted average farm price of all conceivable competing crops in the regions, was employed for all regions in this study. Positive yield response was found in almost all cases.

The price variable was positive in eight cases and negative in six cases. The short run price elasticity ranged from 0.16 to 0.60 and from –0.05 to –0.27 in the case of positive and negative responses respectively found.

Result indicated that the negative price response for nearly half the acreage under groundnut in India, despite its being a commercial crop. Whenever the price coefficient is negative or non-significant, the coefficient of sowing period rainfall turn out to be positive and significant influencing over groundnut acreage in India. The yield component played a more leading and significant role than the price component and where farmers were not responsive to price movements, the sowing period rainfall seems to exert considerable influence in decision-making.

Panpatte's[24] study of acreage response of major kharif and rabi crops in Marathwada region of Maharashtra state was related to the period from 1952-53 to 1971-72. The basic model used was the Nerlovian adjustment model as follows:

$$X_t^{*} = a + b_{pt-1} + CY_{t-1} + gI_{t-1} + hW_t + U_t \quad (1)$$

$$X_t = X_{t-1} = B\left(X_t^{*} - X_{t-1}\right) \quad (2)$$

where,

X_t^{*} = Area that the farmers would plant in the year t,

X_t = Area actually planted in the year t,

P_{t-1} = Relative price of the crop lagged one year t-1,

Y_{t-1} = Relative yield of the crop lagged one year t-1,

gI_{t-1} = Total irrigated area in all crops, lagged one year t-1,

W_t = Rainfall during the sowing time period t,

U_t = error term.

The crops considered for the study were kharif jowar, bajra, cotton, groundnut, rabi jowar, wheat and gram. He studied the impact of price variables along with other variables such as yield, irrigation and rainfall etc., on the farmers acreage allocation decision.

For Marathwada as a whole, he considered cotton, groundnut and bajra as the crops competing with kharif jowar for allocation of kharif acreage. Kharif jowar, cotton and groundnut as competing crops for bajra, kharif jowar, groundnut and bajra as competing for cotton, kharif jowar, cotton and bajra as competing crop for groundnut.

In case of groundnut the value of R^2 varies from 0.54 for Nanded district to 0.93 for Aurangabad district. In most of the cases the coefficient with respect to absolute price was found to be negative but it was significant only in case of Osmanabad district. In case of Aurangabad, Osmanabad, Beed and for Marathwada as a whole, the coefficient of relative price was negative but not significant. The absolute yield has not shown any significant influence in determining the acreage under groundnut. The coefficient of absolute yield was positive in most of the cases but not significant. The coefficient with respect to relative yield was positive and significant in case of Parbhani district.

The coefficient of irrigation was found to be negative and significant in case of Aurangabad and for Marathwada as a whole. It was found positive and significant in case of Osmanabad district.

The coefficient of rainfall during sowing period was found to be positive and significant in case of Beed and for Marathwada as a whole. The coefficient with respect to lagged acreage was positive and significant in case of all the districts of Marathwada and for Marathwada as a whole.

The coefficient of adjustment varies from district to district. The coefficient of adjustment was found negative in case of groundnut crop in Parbhani district of Marathwada region. Long run elasticity was indeterminate due to negative coefficient of adjustment in case of Parbhani district. Short-run and long-run acreage elasticities varied from crop to crop and district to district. Short run acreage elasticities were higher in case of bajra crop in Beed district compared to other districts and other crops in other districts.

He observed that rainfall irrigation had shown significant influence on the farmers decision regarding the allocation of land under kharif jowar, bajra, cotton, groundnut and rabi jowar.

Uma Kapila[25] examined acreage response of groundnut in various states of India viz. Maharashtra, Andhra Pradesh, Tamil Nadu, Gujarat and Karnataka. The study pertains to the period 1951-52 to 1974-75. She employed Nerlovian adjustment model to study the impact of various factors on acreage allocation decisions for groundnut crop. Jowar, bajra and ragi were competing crops.

The model used consisted of two basic structural equations:

$$A_t^{*} = b_0 + b_1 (RP)_{t-1} + b_2 (RY)_{t-1} + b_3 I_t + b_4 (\sigma P)_t + b_5 (\sigma Y)_t + b_6 T + U_t \quad (1)$$

$$A_t - A_{t-1} = c_1 \left(A_t^{*} - A_{t-1}\right) c_2 W_t \quad (2)$$

Where,

A_t = Area under the groundnut crop in thousand hectares,

RP_{t-1} = Ratio of the farm harvest price of groundnut crop to that of competing crop with a lag of one year.,

RY_{t-1} = Ratio of the yield of groundnut crop to that of competing crops with a lag of one year,

I_t = Gross irrigated area/ratio of total irrigated are to total cropped area in the current year.

W_t = Weather referring to rainfall in the millimeters during a specified period of time in the current year.

A_{t-1} = Lagged area/area in period (t-1),

σY_t = Yield variability for year 't' i.e. ratio of coefficient of variation of groundnut yield to coefficient of variation of competing crop yield over the preceding three years It is a measure of technical uncertainty.

σP_t = Price variability for year 't' i.e. ratio of the coefficient of variation of groundnut prices to the coefficient of variation of competing crop prices over the preceding three years, t-3, t-2, t-1. It is a measure of price risk.

T = Trend variable which takes the values 1,2,−n. T variable has been interpreted to represent technology variables in some cases.

She hypothesised that relative price, relative yield, irrigation, relative yield variability, relative price variability, time trend and rainfall were the variables and C_1 was the coefficient of adjustment and C_2 measures the effect of rainfall on acreage change. She had taken relative price and relative yield as factors influencing the relative profitability as well as other explanatory variables such as irrigation and rainfall. Multiple linear regression functions were fitted for estimating the coefficients of acreage response functions for groundnut. The explanatory power of all the selected equations, except of those for Karnataka, was satisfactory. The R^2 ranges between 0.70 and 0.95 for Karnataka it was around 0.52.

In Maharashtra groundnut acreage started declining with negative growth rate of area (-4.85 per cent per annum). In the acreage response functions for the state, she had taken bajra and jowar as the competing crops for groundnut. The relative yield of groundnut with respect to jowar, showed positive and significant coefficients which along with the other variables (relative price, time trend, rainfall and variability factors) explain up to 94 per cent of variation in groundnut acreage. Thus, the decline in groundnut acreage in the western states of Gujarat and Maharashtra could be attributed to strong negative pull of factors other than prices such as technology of competing crop which through its effect on relative yield, reduced the relative profitability of groundnut and diverted groundnut area to the competing crops.

The study brings out clearly that favourable prices by themselves significantly influence acreage. In the western states of Gujarat and Maharashtra as well as in the southern state of Karnataka, in spite of favourable price factor, groundnut could not gain acreage because the competing crops enjoyed a relative technological advantage which out weighted the relative price advantages of groundnut. The uncertainty and risk factors also might have adversely affected groundnut acreage. These factors have turned out to be significant in some of the equations for the western states. However, for southern states, this variable did not turn out to be significant.

In the southern states of Andhra Pradesh and Tamil Nadu, groundnut could gain area due to irrigation which turned out to be the most important factor influencing groundnut acreage. Response of groundnut acreage to irrigation was fairly high in Tamil Nadu; a one per cent increase in total availability of irrigation leads to 0.709 per cent increase in groundnut acreage in the short-run and 1.14 per cent increase in the long-run. In Andhra Pradesh the elasticity of groundnut acreage with respect to gross irrigated area was 2.26 (short run) and 3.35 (long run). The fairly high elasticity of groundnut acreage to irrigation in the southern states of Andhra Pradesh and Tamil Nadu could be responsible for its relative profitability unlike in the western states where irrigation was scarce and groundnut suffered technological disadvantages *vis-à-vis* competing crops.

The relative yield factors have not been as unfavourable in the southern states as in the western states because in the former groundnut could gain from increased availability of irrigation. Since irrigation has been mainly confined to rabi/summer groundnut where yield rate was significantly higher than that in kharif, yield factor has been comparatively favourable in southern states as compared to the western states. The coefficient of relative yield/relative gross income were positive and significant in Andhra Pradesh. It appears that availability of irrigation has increased the relative yield of groundnut and this, combined with favourable relative price, increased the relative profitability of groundnut and thus attracted more area under it.

Garule[26] studied acreage response of kharif jowar by regressing acreage of crop on lagged acreage, lagged real price, lagged yield, pre-sowing rainfall, lagged yield of competing crops, area of competing crops and lagged real price of competing crops. He concluded that area of mung competing crop influenced the area allocation of jowar. He observed non-significant impact of real prices of competing crops on acreage of kharif jowar. The acreage of the crop was influenced by current as well as lagged acreage of competing crop cotton.

Sidhu and Sidhu's[27] study was related to growth and area response of commercial crops in Punjab during the period of 1967-68 to 1985-86. They examined the changes in the composition and growth of commercial crops like cotton, sugarcane, oilseeds, potato, rapeseed, mustard, rice and wheat. They have analysed the factors responsible for determining area under these commercial crops. They used Nerlovian lagged adjustment model. The estimation equation of these crops is as follows:

$$A_t = b_0 + b_1 P_{t-1} + b_2 Y_{t-1} + b_3 \delta_{pt} + b_4 \delta_{yt} + b_5 I_t + b_6 W_t + b_6 W_t + b_7 A_{t-1} + e_t$$

where,

A_t = Acreage under the crop in the year 't' ('000' hectares).

P_{t-1} = Price of the crop in Rs./quintal deflated by the price of substitute crop in the year t-1.

Y_{t-1} = Yield of crop in kg per hectare deflated by the yield of the substitute crop in the year t-1.

δ_{pt} = Standard deviation of preceding three years' prices of the crop.

δ_{yt} = Standard deviation of preceding three years' yield of the crop.

I_t = Gross irrigated area (in thousand hectares),

W_t = Rainfall in the planting period in the year (in mms)

A_{t-1} = Area under the crop in year t-1 (in thousand hectares) and

b_i = the parameters to be estimated.

In their study relative price, relative yield, rainfall, irrigated area, lagged area (t-1), lagged area (t-2), lagged area (t-3) and standard deviations of price and yield were the independent variables.

The study revealed that the area under groundnut was continuously declined. This was due to higher profitability of American Cotton as a competing crop. Irrigated area had also inverse relationship with the area under groundnut. The decline in the area under this crop was gradual as shown by significant positive relationship between the current area and lagged area. According to their opinion the development of new technology in the cereals was responsible for shifting the area from groundnut. Inspite of relative improvements in prices and yield the area under groundnut declined because the crop became economically an inferior one. They found that the relative price of sugarcane and one year lagged area (A_{t-1}) had significant positive influence on the current area. Under sugarcane, whereas two year lagged area (A_{t-2}) had negative significant effect indicating the cyclic-fluctuations in area under sugarcane and production of sugarcane. Other variables were found to be non-significant. This is because sugarcane crop is a three year crop (plant + ratoon + ratoon).

In connection with potato the effect of relative price and lagged area was positive and significant, while the effect of other variables was non-significant which means that price and market situation as it is, is quite favourable for this crop. The relative yield was not contributing significantly to the area under American cotton. The relative price irrigated area and lagged area played significant role in increasing the area under American cotton. The risk of yield, price and rainfall were non-significant. In connection with all other traditional commercial crops, relative price and their stability were the most important factors that influence the area under commercial crops.

Bhagat[28] studied acreage response of individual crops viz. rice, maize, wheat, barley, ragi in terms of acreage elasticity to price and non-price factors in Chotanagpur region of Bihar state. Study pertains to the period 1956-57 to 1976-77. He considered planted area as dependent variable and lagged area, lagged price, lagged yield, lagged gross return per unit of area, current year's pre-sowing rainfall, irrigated area and fertilizer prices as other explanatory variables. Weather was another important variable considered in his acreage equation. Almost all important cereals were grown under rain-fed conditions in the Chotanagpur region. That is why weather factor was important in determining farmers' decision in respect of acreage allocation. Pre-sowing rainfall was taken as a proxy for the weather factor. In this study Nerlovian lagged adjustment model was applied.

He concluded that R^2 were very high in the regression for Chotanagpur region as a whole. The coefficient of the lagged dependent variable for Chotanagpur region were highly significant in all the cases and approximately equal to 1 indicating that almost the same amount of land under the crops has been maintained and found out that insignificant price coefficients in all the cases.

The coefficient of pre-sowing rainfall variable were positive and in most cases significant for winter rice but negative and in most cases significant for maize and ragi. This may be so because maize and ragi require less water than other cereals like autumn rice grown in the kharif season and the heavy rainfall during the sowing season may have adversely affected the area under these crops favouring in their place the other cereals. He found that the price responsiveness of the Chotanagpur farmers was very much constrained by the subsistence characteristics of the crops and their dependence on monsoon. All price and gross return coefficients were insignificant except in a few cases. On the whole, non-price factors appear to be more important than the price factor in influencing farmers' decisions for grading individual crops like rainfall, water, irrigation facilities and traditional considerations are more important than price considerations in the decision of the farmers.

Shreeram Murthy and Eshwari Prasad[29] examined supply response analysis of sesamum in Andhra Pradesh. This study pertains to the period 1969-70 to 1984-85. They developed area and yield models for analysis of supply response. They concluded that sesamum in the current year was significantly influenced by the combined action of lagged area, prices, price-risk and weather. The response of cultivators in terms of area towards price was very low as revealed by price elasticity. The elasticity of weather factor revealed that there was negative association between sowing period rainfall and area.

Sable[30] examined acreage response of commercial crops in Marathwada region of Maharashtra state for the period 1967-68 to 1987-88. He employed Nerlovian partial adjustment lag model for this study by considering sown area as proxy to supply which was the dependent variable and incorporating the lagged relative price, lagged relative yield, gross irrigated area lagged one year, rainfall in sowing period, lagged acreage, gross return per hectare lagged one year and time trend as the independent variables. He estimated the response relationship in the study with the help of regression equation to get the elasticities. He considered cotton and groundnut as the competing crops for sugarcane in Marathwada region, bajra and groundnut as the competing crops for cotton in Marathwada region and in case of groundnut, kharif jowar, bajra and cotton as the competing crops for Marathwada region.

He examined acreage response of three commercial crops viz., sugarcane, cotton and groundnut in Marathwada region In case of groundnut, The regression coefficient of relative price variable was positive for Marathwada as a whole but statistically insignificant. The coefficients of relative price variable are found to be negative in case of Aurangabad, Parbhani, Beed and Osmanabad districts, but the coefficient is only significant at 5 per cent level in case of Aurangabad district. The negative and significant coefficient indicated that the cultivators were not market oriented.

The regression coefficient with respect to relative yield was found to be positive but not significant for Marathwada as a whole. This may be attributed to the farmers' tendency to shift their land area towards cotton, the most competing crop, to get higher yield. The gross irrigated area has exerted a negative bearing on acreage allocation decisions in all cases. This explains that with the expansion of irrigation facilities the area under groundnut had declined. The rainfall received during the sowing period has shown positive impact on acreage allocation of groundnut crop in Marathwada region as a whole. The impact of rainfall was found to be statistically insignificant for Marathwada region as a whole. The lagged acreage variable was an important determinant of acreage under groundnut in Marathwada as a whole. The impact of this variable was found to be statistically significant for Marathwada region. This positive and significant bearing of lagged acreage variable suggests upward trend in area allocation to groundnut in Marathwada region.

Regression coefficients of gross return variable were found negative for Marathwada region. The negative coefficients were not statistically significant. The estimated coefficient of trend variable had negative sign for Marathwada region which indicates downward shift in area under groundnut with the passage of time. The value of R^2 with respect to groundnut for Marathwada region was 0.8840. The short-run acreage elasticity for Marathwada was 0.0817. But the highest short-run acreage elasticity with respect to price indicates greater percentage of area under groundnut affects changes in price of groundnut. The long-run acreage elasticities are generally higher indicating that the larger time is required for acreage adjustments.

The coefficient of adjustment for Marathwada was found to be 0.5232. The results indicate that the farmers in Marathwada region were positively responsive to both price and non-price factors. Among the non-price variables the impact of lagged acreage variable was found to be statistically significant. It shows positive impact on area allocation decisions for groundnut in Marathwada region.

Chandrabanu and Thomas's[31] study was related to supply behaviour of sesamum and groundnut in Kerala state. The study pertains to the period 1961-62 to 1987-88 by using response function of the Nerlovian (lagged adjustment) type model. In their study there were no major competing crops for sesamum and groundnut in the state. The absolute prices were considered for the analysis.

To study the area response the following model was used:

Where,

$$A_t^d = a_0 + a_1 P_t^e + a_2 W_{pt} + a_3 Y_{t-1} + a_4 PR_t + a_5 YR_t + a_6 T + U_t$$

$$A_t - A_{t-1} = B\left(A_t^d - A_{t-1}\right); 0<B<1$$

A_t^d = Desired crop area in period t,

P_t^e = Expected price of crop concerned in period t,

A_t = Area of crop concerned in '000' ha in period t,

Y_{t-1} = Yield of crop concerned in kg/ha in period t-1,

W_{pt} = Total rainfall of two pre-sowing months in mm,

PR_t = Price risk in period 't' measured by the coefficient of variation of price for the preceding three years.

YR_t = Yield risk in period 't' represented by the coefficient of variation of yield for the preceding three years.

T = Time trend,

B = Coefficient of area adjustment.

Acreage response function for sesamum and groundnut estimated in the log linear form was found to be consistently superior to the linear forms both in terms of proper sign and significance of the coefficients and overall explanatory power. At the state level response of sesamum area to price (three year moving average price) appeared positive, though not significant. However, non-price factors like pre-sowing rainfall and lagged yield seemed to exert much more stronger influence on aggregate acreage. Adjustment was found slow, indicating the existence to techno-institutional constraints. Price risk was negatively related with acreage while yield risk was positively related. However, both the coefficients were not significant, indicating that risk factors had no significant influence on acreage allocation decisions.

In the case of groundnut, only the coefficient of lagged area was significant. Expectations regarding price and yield were positively related with acreage, though none of the coefficients were significant in explaining acreage allocation. In general, the elasticity values were very low indicating that changes in price brought about only less than proportional changes in area of these two oilseed crops.

Hedgire[32] analyzed area response of groundnut, safflower and sunflower crops in Marathwada region of Maharashtra state. In case of kharif groundnut, the lagged area of kharif groundnut, current area of kharif jowar and current farm harvest prices of bajra were the most influencing variables. The season or year had negatively significant impact in area allocation. The long-run price elasticity was higher than short-run price elasticity. The area adjustment period was 10 years. In case of safflower, the lagged farm harvest price of safflower and current farm harvest price of gram implied positively significant effect in area allocation of safflower. The area adjustment with previous season was 93 per cent. The short-run price elasticity was lower than long-run price elasticity. The area adjustment period was 1 year. And in case of sunflower, all the variables included under study explained non-significant impact in area allocation. The long run price elasticities were higher than short run price elasticities.

Tripathy and Srinivas Gowda[33] examined growth, instability and area response of groundnut in Orissa during the post-green revolution period 1970-71 to 1989-90. They used Nerlovian lagged adjustment model to examine the area response of groundnut in Orissa. The model has been derived as follows:

$$A_t^* = a_0 + a_1 P_{t-1} + a_2 R_t + a_3 W_t + a_4 I_t + U_t \quad (1)$$

$$A_t - A_{t-1} = C\left(A_t^* - A_{t-1}\right);\ 0 < C < 1 \quad (2)$$

The above two equations are used for estimation where,

A_t^* = Desired area in hectares under the crop in the year t,

P_{t-1} = Farm harvest prices in the period t-1,

A_t = Actual area in hectares under the crop in year t,

A_{t-1} = Actual area in hectares under the crop in year t-1,

R_t = Price risk as measured by the standard deviation of price in the past three years from the period t,

W_t = Rainfall in the planting period in the year t (in mm),

I_t = Gross irrigated area in the year t,

U_t = Error term in the year t,

C = Coefficient of adjustment.

The above equations are used for estimation of:

$$A_t = b_0 + b_1 P_{t-1} + b_2 A_{t-1} + b_3 R_t + b_4 W_t + b_5 I_t + V_t \quad (3)$$

Where,

$$b_0 = a_0C,\ b_1 = a_1C,\ b_2 = (1\text{-}C),\ b_3 = a_2C,\ b_4 = a_3C,\ b_5\ a_4C,\ V_t = U_tC.$$

They found that the lagged area, lagged prices, price risk, and irrigation had positive and significant impact on the area under the groundnut. The effect of rainfall turned out to be negative but statistically non-significant, implying that excess rainfall during the sowing period inhibited area expansion of groundnut. Among different regression coefficients, irrigation had larger value. It indicated that irrigation favourably influenced the area under groundnut in the state during the post-green revolution period.

Singh's[34] study was related to growth and supply response of oilseeds in Uttar Pradesh for the period 1966-67 to 1989-90. He considered rapeseed, mustard, groundnut, linseed and sesamum as main oilseed crops and wheat and paddy were the competing crops for the state. He used the Nerlovian lag adjustment model for supply response study and examined the effectiveness of the additional variables viz., price and yield variability to the area allocation to oilseed crops, the coefficient of variation of price (C_oV_p) and coefficient of variation of yield (C_oV_y) over the preceding three years were included.

The model adopted was as follows:

$$A_t = B_0 + B_1T_1 + B_2I + B_3A_{t-1} + B_4P_{t-1} + B_5Y_{t-1} + B_6P_{ct-1} + B_7C_oV_p + B_8C_0V_y.$$

Where,

- A_t = Area under oilseed crop for current year,
- T_1 = Time trend variable,
- I = Current year's irrigated area under oilseed crop,
- A_{t-1} = Area under oilseed crop lagged by one year,
- P_{t-1} = Farm harvest price of oilseed crop lagged by one year,
- Y_{t-1} = Yield of oilseed crop lagged by one year,
- P_{ct-1} = Farm harvest price of competing crop lagged by one year,
- C_oV_p = Coefficient of variation of price of oilseed crop (price risk)

C_oV_y = Coefficient of variation of yield of oilseed crop (yield risk),

B_i = Regression coefficient (i = 0, 1, ...8.)

In connection with Rapeseed-Mustard area response, wheat crop was as competing crop of rape-seed-mustard in the state. The regression coefficient of time variable was negatively significant. Irrigated area under this crop and yield risk depicted significant influence with a positive trend. Farmers were aware about yield risk of rapeseed-mustard crop. These variables significantly contribute to current year's area allocation under this crop. These variables significantly contributed to current years area allocation under this crop. As far as price risk is concerned, it is not significant in decision-making regarding area allocation under rapeseed-mustard.

Regarding groundnut area response, paddy has been considered as major competing crop for groundnut in Uttar Pradesh. The factors of time variable, current year's irrigated area, lagged area, lagged price, lagged yield, lagged price of competing crop, price and yield risk had influenced the farmer's land allocation of decision regarding current year area under groundnut. The coefficients of current year irrigated area, lagged area, lagged yield and yield risk were significant with positive sign, whereas lagged price of groundnut and price risk were significant with negative sign. It showed that the spread of irrigation facility has affected groundnut cultivation. Farmers preferred to allocate more irrigated area under groundnut cultivation. The previous year area under groundnut and yield risk affects farmers during current year area allocation decision-making. The previous year price of groundnut and price risk have an important role in decision-making.

Singh examined area response of linseed crop assuming wheat was a competing crop for linseed in Uttar Pradesh. He explained that all explanatory variables jointly affect farmers' land allocation decision by 92.90 per cent. The coefficients of current year's irrigated area (I) and previous year's yield of crop concerned as observed were found significant with a positive sign. It revealed that the farmers preferred to allocate more irrigated area for linseed crop when irrigation facility was available. It was also found that a unit increase in yield of linseed crop cultivation had a positive effect on farmers' decision on linseed area. Farmers in the state were quite aware about yield risk of crop concerned.

Lastly, he explained that paddy was competing crop for sesamum in Uttar Pradesh. The coefficient of multiple determination (R^2) was found 59.20 per cent. It indicates that the explanatory variables i.e. time variable, current irrigated area, lagged area, lagged price, lagged yield, lagged price of competing crop, price and yield risk jointly influenced by 59.20 per cent in land allocation decision. The coefficient of time variable recorded a significant effect with a negative sign. While last year's price of sesamum showed positive significance. It depicted that the farmers in Uttar Pradesh were more responsive towards price of concerned crop, and this variable significantly influenced the farmers decision in current year's area allocation for sesamum cultivation. It has also been observed that the non-price factors have influenced the decision-making process more than the price factor. Farmers are preferring to allocate more irrigated area under oilseed crops in Uttar Pradesh.

Dixit, Hiremath and Raj Vir Singh[35] analysed supply response of groundnut crop in Karnataka state. Their study was concerned with the period from 1975-76 to 1992-93 pertaining to the hectarage, yield, production and post-harvest prices of groundnut and its competing crops. Jowar, cotton, chilli and ragi were competing crops to the groundnut. Nerlovian expectation and adjustment model was used in this study. Price and non-price factors were as variables. They concluded that the pre-sowing months rainfall registered its statistically significant influence in all the districts barring the aggregative function for Karnataka. The positive impact of rainfall factor on groundnut production revealed its pivotal role in acreage allocation decision. It was conceivable that a favourable rainfall in the pre-sowing months would enable the farmers to allocate more area to groundnut than before in anticipation of harvesting good crop. Similar conclusions emerged from the computed elasticity coefficients with respect to the pre-sowing months' rainfall is the negative relation of rainfall with an acreage allocation at the state level could be attributed to the aggregation bias.

Shakuntala Gupta[36] examined the acreage response of major oilseed crops in Punjab to changes in their relative price. This study primarily focuses on the allocation of land resources and covers the period of 1966-67 to 1995-96. This study proposed to examine the impact of various price and non-price variables on acreage of both groundnut and rapeseed-mustard at the state level and district levels as acreage under these crops has fallen down at the state level and as well as at the district level. In this study wheat was the only major competing crop for rapeseed-mustard, whereas in Kharif, paddy was selected as the competing crop for groundnut.

The estimating equation of the Nerlovian model was as follows:

$$A_t = b_0 + b_1 P_{t-1} + b_2 Y_{t-1} + b_3 \sigma_{pt} + b_4 \sigma_{yt}\, b_5 I_t + b_6 W_t + b_7 A_{t-1} + V_t$$

where,

A_t = Planted /actual acreage of the concerned oilseed crop in the year t,

A_{t-1} = Planted acreage of the concerned oilseed crop in the year t-1.

P_{t-1} = Harvest prices of the concerned oilseed crop relative to the price of its competing crops in the year t-1,

Y_{t-1} = Yield of the concerned oilseed crop relative to yield of the competing crop in the year t-1,

σ_{pt} = Standard deviation of preceding three years' prices of concerned oilseed crop relative to that of the competing crop,

σ_{yt} = Standard deviation of preceding three years' yield rates of concerned oilseed crop relative to that of the competing crop,

I_t = Gross irrigated area under all crops in a season,

W_t = Rainfall during a specified period, defined as an average of rain received in June-July months in case of groundnut crop and October-November months in the caser of rapeseed-mustard in the year t (in millimeters),

V_t = Error term.

b = denote the respective variables slope coefficients.

Multiple linear regression function using ordinary least-squares method was fitted to calculate coefficient of acreage response functions. In order to examine the relative importance of variables, influencing the farmers acreage allocation behaviour, acreage elasticities were calculated for price and non-price variables like yield, risk in yield, irrigation, rainfall and lagged area. Both short-run and long-run elasticities of price and non-price variables were made for Bhatinda and Sangrur districts separately and Punjab state as a whole. Lagged area, relative price, relative yield, relative risk in yield, relative risk in price, gross irrigated area and rainfall were the variables. The lagged area variable, irrespective of time and of the measurement level, state or district, showed positive significant coefficient in the vicinity of unit value. The relative yield, the second explanatory variable in the formulation, bears the expected positive sign. The value of coefficient was statistically significant though at the district level, the value of coefficient was 4.31. This suggests that the decisions to put area under groundnut crop were significantly influenced by improvement in relative yield in Punjab state. The coefficient relating to relative price bears positive sign at the state level where economic considerations have been observed to be playing a decisive role in area allocation under the groundnut crop. Computed elasticities lend support to the farmers positive response to price changes in acreage allocation. Relative risk in yield variable, bears negative coefficients. This was statistically significant too at the state level. Elasticity estimates also revealed the same picture. As the relative risk in yield increases, they tend to reduce the area under the crop though they do not bother much about relative risk in price at the state level.

Since groundnut is an unirrigated crop, as the irrigated area increases, the area under it is found to decrease. This fact is amply portrayed by the negative significant value of gross irrigated area variable. This suggests that groundnut cultivation in Punjab was confined to marginal lands which were not very prone to irrigation facilities. Rainfall, a direct source of water to the plant, bears a positive sign but the value of the coefficients in this case was not statistically significant. To sum up the acreage allocation under groundnut crop in Punjab state was generally governed by economic, technological and structural reasons.

In case of rapeseed-mustard acreage-behaviour does not differ substantially from that of groundnut. The role of the technology of the substitutes in displacing area under rapeseed-mustard was duly brought out by the negative significant coefficient of gross irrigated area at the state (-0.019) as well as at the district level (-0.044). Elasticity of irrigation also accounts for a greater amount of change in acreage and imprinted its negative influence. In the behaviour of variable of relative risk in yield in the district and in the state acreage under rapeseed mustard was primarily confined to marginal land. The coefficient of yield was found to be positive as well as significant too. This suggests that decisions to put area under rapeseed-mustard were significantly influenced by improvement in the yield both in the state and at the district level. In respect of market response as relative price becomes favourable, farmers go in for more rapeseed-mustard, value of both the coefficients is positive but significant at the state level only. Similarly, as the relative risk in price increases, they tend to reduce the area under the crop, both the coefficients bear negative sign but not statistically significant at the district level only. The negative coefficient value of variable clearly reveals that groundnut as well as rapeseed-mustard cultivation in Punjab, particularly in Bhatinda and Sangrur districts respectively, was probably confined to marginal lands which are not very prone to irrigation facilities.

In case of groundnut crop, it was found that elasticities with respect to irrigation and lagged area were much higher than price elasticity as well as yield elasticity of acreage, whereas risk variables elasticities under low response category and these variables imprinted their negative influence on acreage under groundnut. In fact, both at the state and district level, irrigated and lagged area appear to be significant causing great influence on the farmers' decision-making process. The long run price elasticity was high but short run elasticity of relative price variable was not high as compared to non-price variables influencing the acreage under groundnut compared to price variables. The result indicates that farmers have constraint in adjusting their acreage to the desired level in the short-run and they will take a long time to adjust.

In case of rapeseed-mustard it was found that elasticities with respect to irrigation and relative price, the elasticity of irrigation accounts for a greater amount of change in acreage than price elasticity and this variable imprinted its negative influence on acreage under rapeseed-mustard whereas price registered positive influence on rapeseed-mustard acreage. The study found that relative yield, relative price and irrigation are the major factors affecting acreage allocation decisions. This fact would require improvement in their yield, removal of productivity constrains, effective price support as well as irrigation coverage to oilseed crops, as there is a sufficient scope for the adjustment of area under oilseed crops in Punjab.

The literature reviewed in this chapter indicated that the farmer is a control point who makes final decisions regarding the allocation of land to different crops. It is generally hypothesized that an acreage in the price induces the farmers to increase the area under cultivation to the best possible extent, to apply more inputs to produce the maximum possible quantum using the best available technology.A considerable amount of literature has appeared on acreage response. At present two different views on acreage response prevail one view is that the farmers in underdeveloped countries are not responsive to the changes in relative price and the other view is that they are no less responsive than their counterparts of the developed countries like the United States.Literature reviewed in this chapter further indicated that change in relative price,relative yield,rainfall had shown significant influnces on the farmers decision reagarding the allocation of land under considered crops.

NOTES AND REFERENCES

1. Padwal, R. N. (1954), Production and Price Trends of Major Oilseeds in India, *Indian Journal of Agricultural Economics*, 9(56).
2. Nerlove Marc (1958), "Dynamics of Supply: Estimation of Farmers Response to Price", John Hopkins Press, Baltimore.
3. Mishra, B. and Sinha, S. P. (1958), "Agriculture and its Term of Trade with Special Reference to India", *Indian Journal of Agricultural Economics.*
4. Rajkrishna (1963), "Farm Supply Response in India-Pakistan: A Case Study of the Punjab Region", *The Economic Journal*, Vol. LXXIII, No. 291, September.
5. John, P. V. (1965), "Some Aspects of the Structure of Indian Agricultural Economy, 1947-48 to 1961-62", Asia Publishing House, New Delhi.
6. Kamala Devi, P. and Rajgopalan, R. (1965), "Price and Acreage Response", A Case Study of Groundnut in North Arcot District, Madras State, *Indian Journal of Agricultural Economics*, Vol. XXI, No. 1.
7. Dharam Narain (1965), "Impact of Price Movement on Areas Under Selected Crops in India, 1900-1939", Cambridge University Press.
8. Rajgopalan, V. (1967), "Supply Response for Irrigated Crops in Madras State, India", Ph.D. Dissertation, University of Tennessee.
9. Behraman, J. R. (1968), "Supply Response in Underdeveloped Agriculture", A Case Study of Four Major Annual Crops in Thailand (1937-163), North Holland Publishing Company, Amsterdam.
10. Reddy, J. M. (1970), "Estimation of Farmers Supply-A Caser Study of Groundnut", *Indian Journal of Agricultural Economics*, Vol. XXV, No. 4.
11. Kaul, J. L. and Sidhu D. S. (1971), "Acreage Response to Prices for Major Crops in Punjab"-An Economic Study, *Indian Journal of Agricultural Economics*, Vol. XXVI, No. 4.
12. Misra, V. N. and R. Radhakrishnan (1971), "Economic Models for Acreage Response-A Study of Groundnut and Bajra in Gujarat", *Anvesak*, Vol. 1, No. 2.
13. Sahay, K. K. (1971), "Acreage Response of Groundnut in Major Groundnut Zones: A Case Study in Dynamics of Supply, *Agricultural Situation in India*, Vol. XXVI, No. 7.

14. Parikh, A. (1972), "Market Responsiveness of Peasant Cultivators: Some Evidence From Pre-War India", *Journal of Development Studies*, No. 8.

15. Madhavan, M. C. (1972), "Acreage Response of Indian Farmers: A Case Study of Tamil Nadu", *Indian Journal of Agricultural Economics*, Vol. XXVII, No. 1.

16. Mohapatra, B. K. (1972), "The Impact of Price Changes on Areas Under Selected Crops in Orissa", *Asian Economic Review*, Vol. 14.

17. Gupta, R. P. (1973), Agricultural Prices in Backward Economy, National Publishing House, Delhi.

18. Singh, R. P., Singh Parmatma and Raj K.N. (1973), "Acreage Response to Rainfall, New Farm Technology and Price in Haryana", *Indian Journal of Agricultural Economics*, Vol. 54, No. 4.

19. Askari Hussain and Cummings, J. T. (1974), "Output Price Response in Agriculture: An Evaluation", *Indian Journal of Agricultural Economics*, Vol. XXXIX, No. 2.

20. Acharya, S. S. and Agrawal, N. L. (1974), "Factors Affecting Supply and Growth Rates of Commercial Crops in Rajasthan", *Indian Journal of Agricultural Economics*, Vol. XXIX, No. 3.

21. Acharya, S. S. and Bhatiya, S. (1974), "Acreage Response to Price, Yield, Rainfall Changes in Rajasthan", *Agricultural Situation in India*, Vol. XXVIII, No. 4.

22. Sidhu, D. S. and Prakash Mehta (1975), "Response of Farmers to Prices–A Case Study of Potato and Groundnut in Punjab", *Economic Affairs*, Vol. XXIX, No. 6.

23. Jhala, M. L. (1979), "Farmers Response to Economic Incentives: An Analysis of Inter-Regional Groundnut Supply Response in India", *Indian Journal of Agricultural Economics*, Vol. XXXIV, No. 1.

24. Panpatte, B. P. (1980), Some Structural Aspects of Agricultural Development in Marathwada, Unpublished Ph.D. Thesis, submitted to Marathwada University, Aurangabad, June.

25. Uma Kapila (1982), "Acreage Response of Groundnut–A State Level Study", Oil Seeds Economy of India, Agricultural Publishing Academy, New Delhi.

26. Garule, J. H. (1984), "Acreage Response of Kharif Jowar in Parbhani District of Maharashtra State", M.Sc. (Agri.), Thesis Submitted to Marathwada Agriculture University, Parbhani.

27. Sidhu, J. S. and Sidhu, R. S. (1988), "Growth and Area Response of Commercial Crops in Punjab: Some Policy Issues", *Indian Journal of Agricultural Economics,* Vol. XLIII, No. 3.

28. Bhagat, L. N. (1989), "Supply Response in Backward Agriculture–An Economic Study of Chotanagpur Region (Bihar), Concept Publishing Company, New Delhi.

29. Shreeram Murthy, C. and Eshwari Prasad (1990), "Supply Response Analysis of Sesamum in Andhra Pradesh", *Agricultural Situation in India,* Vol. XLIV, No. 11.

30. Sable, M. S. (1991), "Supply Response of Some Agricultural Commodities in Marathwada", Unpublished Ph.D. Thesis, Submitted to Marathwada University, Aurangabad, August.

31. Chandrabanu, P. and Thomas, E. K. (1991), "Supply Behaviour of Sesamum and Groundnut in Kerala", *Agricultural Situation in India,* Vol. XLVI, No. 5, August.

32. Hedgire, D. N. (1992), "A Study of Production, Marketing and Processing of Major Edible Oilseeds in Marathwada–An Economic Analysis", Unpublished Ph.D. Thesis, Submitted to M.A. University, Parbhani.

33. Tripathy, S. and Srinivasa Gowda, M. V. (1993), "An Analysis of Growth, Instability and Area Response of Groundnut in Orissa", *Indian Journal of Agricultural Economics,* Vol. 48, No. 3.

34. Singh, O. P. (1998), "Growth and Response of Oilseeds in Uttar Pradesh", *Indian Journal of Agricultural Economics,* Vol. LV, No. 1.

35. Dixit, P. K., Hiremath, K. C. and Rajvir Singh (1998), "Production Behaviour of Groundnut Farmers in Karnataka: A Dynamic Supply Response Analysis", *Indian Journal of Agricultural Economics,* Vol. 53, No. 2.

36. Shakuntala Gupta (2000), "Acreage Response of Major Oilseed Crops in Punjab", *Agricultural Situation in India,* Vol. LVI, No. 12, March.

3 Agricultural Economy of Marathwada Region

Aurangabad division of Maharashtra state known as 'Marathwada', consisted of eight districts, viz. Aurangabad, Parbhani, Beed, Nanded, Osmanabad, Latur and Hingoli. Till 1981, the region consisted only five districts, but in May 1981 Aurangabad district was divided into Aurangabad and Jalna districts and in August 1982, the Osmanabad district was divided into Osmanabad and Latur districts. And in May 1999 Parbhani district divided into Parbhani and Hingoli districts. Although Marathwada region has now eight districts, even then the total geographical area of region remained unchanged.

The entire region formed as a part of the ex-princely state of the Nizam, then known as Hyderabad state. The Hyderabad state was merged into the Indian Union in 1949. In 1956, with the reorganization of states, the Marathwada region was joined to the bilingual Bombay state which consisted of the present Maharashtra and Gujarat states. It has been one of the four administrative divisions of Maharashtra state since 1st May, 1960.

Geographically, the region is situated between 17°35¢N and 20°40¢N latitudes and 74°40¢E and 78°15¢ longitudes. The region is bounded by Jalgaon, Buldhana and Akola districts in the north, by Nasik and Ahmednagar districts in the west, Solapur district in the south and Andhra Pradesh state in the east. The total geographical area of the region is 64,434 square kilometers, which is 12 per cent of Maharashtra state. The districtwise area is mentioned as below: Aurangabad 10077 sq. kms, Jalna 7726 sq. kms, Beed 10,686 sq. kms, Parbhani 10,972 sq. kms, Nanded 10331 sq. kms, Osmanabad 7485 sq. kms, Latur 7157 sq. kms and Hingoli 5521 sq. kms. The population of the region according to 2001 census is 155.89 lakhs[1] and about 75.41 per cent of total population lives in rural areas and 24.59 per cent in urban areas.

Agriculture is one of the most important sectors of this region. In the year 2001, 76.72 per cent of the workers in Marathwada region were engaged in agriculture[2], which reveals that the major portion of the population of this region depends on agriculture. Some basic agro-economic features of Marathwada region are as follows:

THE HILL RANGES

There are two major hill ranges in Marathwada region, the Satmala range in the north and the Balaghat range in the south. The northern hill ranges are the extension of Satmala ranges from Nasik and include famous projections like Ajanta, Chowtra and Jostna hills. The Daulatabad hill range runs through the center of Aurangabad district eastward demarcating the fertile valleys of Girja and Purna in north and Dudhana on the south[3]. The Satmala or Ajanta range starts at the north-western corner and runs close and parallel to the northern border of the Marathwada region up to the Ajanta acting as the watershed between the Tapi valley in the north and the Godavari valley in the south. The river Godavari flows along the southern boundary of the district and enters the Nanded district. The river Dudhana meets the Purna river near Parbhani town and then joins the Godavari river.

Another range of hills which is a continuation of hill ranges in northern Parbhani extends from east to west in the Nanded district and the separate valleys of Penganga in the north and the Godavari and its tributaries in the south. Except the northern portion and some part of Nanded and Kandhar talukas which are hilly, the rest of the district has almost plain land[4]. Another branch of the Satmala, known as the Jalna hill range, runs through the center of the Aurangabad district (now through Jalna district). The Balaghat range runs through the center of the Beed district acting as the water-shed between the Godavari basin in the north and the Bhima basin in the south. Besides these two ranges running east-west in the region, there is a third range i.e. Thanwari range which runs north-west to south-east in the Nanded district. From the north-west boundary of Osmanabad district runs a range of hills which divides into two parts, plateau on the north-east and the low land in the remaining area. The river Manjara flowing along the northern boundary of the Osmanabad and Latur districts runs towards the south and enters the Bidar district of Karnataka state. The hills in the Marathwada region are of moderate height and have terrace like appearance with flattened summits.

THE FORESTS

The area of forest in the region is very small[5] i.e. 3.18 per cent of geographical area of the region in the year 1999-2000 and it is concentrated in Kinwat taluka of Nanded district. Teak, bamboo, gum and grass are the main forest products of this area. The forest patches in Aurangabad district yield some timber and small quantity of sandal wood. The scattered forest patches in Beed and Parbhani districts produce mainly wood which is mostly used directly as fuel or for making coal. The area under forest in Osmanabad district is very negligible, i.e. 0.14 per cent of geographical area of the district[6] in the year 1999-2000.

To improve the climate and rainfall in the Marathwada region, the area under the forest must be increased by implementing various forest department's schemes in the region and create new forests on near about 1/3 of the region's area.

THE MINERALS

Marathwada is almost devoid of mineral wealth. The production of minerals is not considerable in Aurangabad district. Jaspers, Agata, Carnelian and some rock crystals are found in Aurangabad district. Nodular lime stone, basalt and granite occur all over the Aurangabad and Jalna districts.

The production of minerals in Beed district is also not sufficient and considerable. Basalt, lime-stone and silicon are the minerals found and used for building purpose. The position of Osmanabad and Latur districts is the same. Production of minerals in Parbhani district is negligible. Production of limestone is found in Nanded district but quality of limestone is not uniform all over the district. Metal, silicon and limestone type of minerals are found in Nanded district and it is used for building purpose.

SOILS

Most of the region is covered with black cotton soil or 'Regur' derived from the Deccan trap volcanic rock. However, the soils vary widely in texture and depth. The deep black soils are found along the river banks of Godavari, Manjra, Purna, Dudhana and Penganga and their tributaries, and the soil is capable of retaining moisture. Bumper rabi crops are grown in these areas in the year of favourable rainfall. The soils are, however, coarse, shallow and relatively poor along hill slopes, and at the foot of the hills.

The major portion of the region is covered by medium black soil. The medium and deep soil in the region is rich in plant nutrients and can support good kharif or rabi crops like jowar, bajra, wheat, pulses, cotton and groundnut. Medium and light soils with less depth is found on the hill tops or along the hill slopes. It is suitable only for kharif crops because it is relatively weak and not retentive of moisture. The area belonging to various classes in different districts is presented in Table 1.

It is clear from the table 3.1 that the nature of soils is not uniform in the Marathwada region. Major portion of the region is covered by medium black soil i.e. 64.79 per cent. The deep black soil covers about 12.95 per cent of the region and the coarse and shallow soils cover about 22.26 per cent portion of the Marathwada region.

The deep black soil covers about 12.95 per cent of Marathwada region. But, the percentage of deep black soils is highest in the Nanded district, i.e. 24.01, followed by Parbhani district i.e. 20.88 per cent, Beed i.e. 12.26 per cent,Jalna i.e. 10.62 per cent, Aurangabad i.e. 8.14 per cent and lowest in the Osmanabad and Latur districts i.e. 3.95 per cent and 3.89 per cent respectively.

Table 1
Districtwise Classification of Soils in Marathwada Region during the year 2001

(Fig. in '00' hectares)

	Type of Soil			
District	*Deep black more than 36" depth*	*Medium black between 9" to 36" depth*	*Coarse and shallow below 9" depth*	*Total*
Aurangabad	82	587	339	1008
	(8.14)	(58.23)	(33.63)	(100)
Jalna	82	461	229	772
	(10.62)	(59.72)	(29.66)	(100)
Beed	137	726	254	1117
	(12.26)	(65.00)	(22.74)	(100)
Osmanabad	29	619	87	735
	(3.95)	(84.21)	(11.84)	(100)
Latur	26	563	79	668
	(3.89)	(84.28)	(11.83)	(100)
Parbhani	229	715	153	1097
	(20.88)	(65.17)	(13.95)	(100)
Nanded	248	495	290	1033
	(24.01)	(47.92)	(28.07)	(100)
Marathwada	833	4166	1431	6430
	(12.95)	(64.79)	(22.26)	(100)

Note: Figures in brackets indicate percentages.
Source: Data supplied by the Zilla Parishad.

The deep black soil covers about 12.95 per cent of Marathwada region. But, the percentage of deep black soils is highest in the Nanded district i.e. 24.01 per cent and lowest in Latur district i.e 3.89 per cent.The medium black soil covers about 64.79 per cent portion of the Marathwada region. But, the percentage of medium black soil is the highest in the Latur district i.e. 84.28 per cent and the lowest in the Nanded district i.e. 47.92 per cent.The coarse and shallow soil cover about 22.26

per cent portion of the Marathwada region. But the percentage of coarse and shallow soils is the highest in the Aurangabad district i.e. 33.63 per cent and lowest in the Latur districts i.e 11.83 per cent.

Black soil is suitable for kharif and rabi crops like jowar, bajra, wheat, cotton, pulses and groundnut. Deep black soil is also found in Bhokardan taluka and some parts of Vaijapur, Gangapur, Paithan, Sillod and Aurangabad talukas and Khultabad Mahal in Aurangabad district; Ambad, Partur and Jalna talukas in Jalna district; Georai and Manjlegaon talukas in Beed district; Biloli, Naigaon, Nanded, Kinwat and Kandhar talukas in Nanded district. The soil of superior quality is also found in Degloor and Hadgaon talukas in Nanded district. Deep black and light soils are found in Osmanabad, Kallam and Omerga talukas in Osmanabad district. Black cotton soil is found in Nilanga taluka of Latur district. Medium and light soils are found in Ahmedpur and Udgir talukas and heavy deep black soil and light soil is found in parts of Latur and Ausa talukas in Latur district. Medium and heavy black soil is found in Parbhani, Basmat, and Pathri talukas and in some parts of Jintur and Gangakhed talukas in Parbhani district and also in Kalamnuri and Akhada Balapur and Hingoli talukas of Hingoli district.

AGRO-CLIMATIC ZONES AND RAINFALL

The state of Maharashtra has been divided into nine broad agro-climatic zones taking into consideration the factors like soils, crops, altitude, rainfall, temperature etc[8]. Marathwada region falls under three zones viz. (i). scarcity zone (ii) assured rainfall zone and (iii) moderate to moderately high rainfall zone. The scarcity zone (700 mm) includes western trap of Beed and Aurangabad districts and Bhoom and Paranda talukas of Osmanabad district. The assured rainfall zone (700-900 mm) covers western half of Parbhani district, southern Nanded and the remaining parts of Aurangabad, Beed and Osmanabad districts. Moderate to moderately high rainfall zone (above 900 mm) covers the eastern part of Parbhani district and northern portion of Nanded.

The average annual rainfall is not uniform in all the districts of Marathwada region. Marathwada region receives an average rainfall of 827 mm. Nanded and Parbhani districts are considered to be the areas with assured rainfall. These districts receive an average annual rainfall between 982 mm and 968 mm respectively. The Table 2 shows that the average annual rainfall is the highest (982 mms) in Nanded district and it is the lowest (719 mms) in Beed district.

RIVER SYSTEM

There are three important river basins in the Marathwada region, viz. the Godavari, the Purna and the Penganga. The major portion of Marathwada region lies in the Godavari basin. Therefore, Godavari is the most important river in the Marathwada region. 'Purna' which is the biggest tributary to Godavari, is the second important river in the region. The river Godavari enters the region through Aurangabad district, runs on its southern boundary, separating it from Beed district, flows through Jalna, Parbhani and Nanded districts to enter Andhra Pradesh. The Purna basin comprises of the whole taluka of Sillod of the Aurangabad district, Jafferabad taluka of the Jalna district, northern part of Kannad taluka, eastern part of Khultabad taluka of the Aurangabad district and Bhokardan taluka of the Jalna district excluding small north-eastern portion forming a salient feature between Jalgaon and Buldhana districts. Purna enters the region from the north-west corner of Parbhani district and flows south-east direction to fall into the Godavari about 15 miles upstream from Nanded district. The Penganga after collecting the water of the southern belts of the Buldhana and Akola districts, drains the north-eastern margin with the aid of its tributary the Kayadhu river.

The other important rivers are Shivna, Dheku, and Kham in Aurangabad district, Manjra, Bindusara and Sindphana in Beed district, Penganga, Manyad, Ashna and Lendi in Nanded district, Terna and Manjra in Osmanabad district, Manjra in Latur district and Kayadhu, Karpara and Dudhana in Parbhani district, and Kayadhu, Purna, Penganga and Asana in Hingoli district.

Table 2
Districtwise Annual Rainfall in Marathwada Region

(Rainfall in mm)

Year	Aurangabad Actual rainfall	Jalna Actual rainfall	Beed Actual rainfall	Osmanabad Actual rainfall	Latur Actual rainfall	Parbhani Actual rainfall	Nanded Actual rainfall	Marathwada Actual rainfall
1980-81	617.2	925.1	608	767	893.2	922	967	814.21
	(85.13)	(119.99)	(84.56)	(86.45)	(107.36)	(95.25)	(98.47)	(98.45)
1981-82	738	850.8	830.6	819.8	850.8	861.5	1004	850.79
	(101.79)	(110.35)	(115.92)	(104.83)	(102.26)	(89.00)	(103.24)	(102.88)
1982-83	478.4	756	507.6	719.6	670.1	608.4	905.7	663.69
	(65.99)	(98.05)	(70.60)	(92.02)	(80.54)	(62.85)	(92.23)	(80.25)
1983-84	1033.1	1224.9	968.5	1077.2	1224.9	1449.1	1596.3	1224.86
	(142.50)	(158.87)	(134.70)	(137.75)	(147.22)	(149.70)	(162.56)	(148.11)
1984-85	374.8	522.8	602.4	430.5	522.8	746.8	459.1	522.74
	(51.70)	(67.81)	(83.78)	(55.05)	(62.84)	(77.15)	(46.75)	(63.21)
1985-86	579	579	483.5	596.7	596.7	774	493.1	586.00
	(79.86)	(76.00)	(67.25)	(76.30)	(71.72)	(79.96)	(50.21)	(70.86)
1986-87	582	499	481.8	486.1	486.1	707	620.5	551.79
	(80.28)	(64.72)	(67.01)	(62.16)	(58.43)	(73.04)	(63.19)	(66.72)

Contd...

(Rainfall in mm)

Year	Districts							
	Aurangabad	*Jalna*	*Beed*	*Osmanabad*	*Latur*	*Parbhani*	*Nanded*	*Marathwada*
	Actual rainfall	*Actual rainfall*	*Actual rainfall*	*Actual rainfall*	*Actual rainfall*	*Actual rainfall*	*Actual rainfall*	*Actual rainfall*
1987-88	745.6	797.8	779.6	781.1	775	845	821	792.16
	(102.84)	(103.48)	(108.43)	(99.88)	(93.15)	(87.29)	(83.60)	(95.79)
1988-89	1007.8	1082	1050	1295.9	1247	1465.6	1815.3	1280.51
	(139)	(140.34)	(146.04)	(165.72)	(149.88)	(151.30)	(184.85)	(154.84)
1989-90	819	701	998.7	1005.3	1004	1112.7	1282.8	989.07
	(112.97)	(90.92)	(138.90)	(128.55)	(120.67)	(114.95)	(130.63)	(119.60)
1990-91	1090.2	889	984.3	1115.1	1191	1386.2	1416.5	1153.19
	(150.37)	(115.30)	(136.90)	(142.60)	(143.15)	(143.20)	(144.25)	(139.44)
1991-92	748	609	405.7	474.1	470	605.8	717.3	575.70
	(103.17)	(78.99)	(56.43)	(60.63)	(56.49)	(62.58)	(73.04)	(69.61)
1992-93	675	773	627.7	587.2	662	765.2	810	700.01
	(93.10)	(100.26)	(87.30)	(75.09)	(79.57)	(79.05)	(82.49)	(84.64)
1993-94	772	791	758	794.9	713	798.6	823.4	778.70
	(106.48)	(102.59)	(105.42)	(101.65)	(59.32)	(82.50)	(93.85)	(94.16)
1994-95	691	591	363.9	470.2	567	1320.5	713	673.87
	(95.31)	(76.65)	(50.61)	(60.13)	(68.15)	(136.42)	(72.61)	(81.48)
1995-96	513	497	602.9	664	873	1109	1152.8	773.10
	(70.76)	(64.46)	(83.85)	(84.91)	(1C4.93)	(114.57)	(117.29)	(93.48)

Contd...

(Rainfall in mm)

Year	Districts							
	Aurangabad	Jalna	Beed	Osmanabad	Latur	Parbhani	Nanded	Marathwada
	Actual rainfall	Actual rainfall	Actual rainfall	Actual rainfall	Actual rainfall	Actual rainfall	Actual rainfall	Actual rainfall
1996-97	740	688	795	866	1000	954.9	944	856.41
	(102.07)	(89.23)	(110.57)	(110.74)	(120.19)	(98.65)	(96.13)	(103.56)
1997-98	717	770	718	703	766	885.1	967.6	789.53
	(98.90)	(99.87)	(99.86)	(89.90)	(92.07)	(91.44)	(98.53)	(95.47)
1998-99	891.9	1166.6	1139.9	1261.5	1253.3	1283	1159.9	1165.16
	(123.02)	(151.31)	(157.71)	(161.32)	(150.64)	(132.54)	(118.12)	(140.89)
1999-00	685	706	681.1	717	884	930.3	978	797.34
	(94.48)	(91.57)	(94.73)	(91.69)	(106.25)	(96.11)	(99.59)	(96.41)
Annual average rainfall	725	771	719	782	832	968	982	827
	(100)	(100)	(100)	(100)	(100)	(100)	(100)	(100)

Note: (i) Figures for newly created Hingoli District are included in undivided Parbhani district
(ii) Figures in brackets indicate percentage of rainfall.

Sources: 1. Season and Crop Report, Government of Maharashtra, 1979-80 to 1995-96, 1997-98, 1989-99,
2. Epitome of Agriculture in Maharashtra Part-I and II, 1984-85, 1987-88, 1988-89, 1994-95 to 1998-99, 2001-2002.
3. Socio-Economic Review and Districts Statistical Abstract of Aurangabad, Jalna, Beed, Parbhani, Osmanabad, Latur and Nanded Districts for the year 1979-80 to 1999-2000.

The following Table 3 shows the three important rivers, which flow through Marathwada region. The length they travel and the area they cover are given in the table below.

Table 3
The Length, Catchment Area and Annual Flow of Water of Three Important Rivers in Marathwada Region

River	*Length (kms)*	*Catchment area (km²)*	*Annual flow (Million cub. mtrs.)*
Godavari	1465	151083	105000
Purna	373	15579	2830
Manjra	724	30844	7640

Source: Marathwada At A Glance: A Case Study Conducted by Marathwada Agricultural University, Parbhani, 1996.

LAND USE PATTERN

Districtwise land use pattern in the Marathwada region is given in the Table 4. In the Marathwada region out of the geographical area of 64,43,400 hectares, about 3.51 per cent (2,26,100 ha) has been classified as forest area under land utilization statistics of the year 1999-2000.

The major portion of the total geographical area i.e. 71.85 per cent comes under net area sown. Barren and uncultivable land is 1.78 per cent and cultivable waste land is 3.26 per cent. Current fallows and other fallows land is 9.62 per cent. Permanent pastures and grazing land and land under miscellaneous trees, crops and groves account for 4.31 per cent and land put to non-agriculture use is 3.68 per cent.

From the Table 4 it is clear that the percentage of area under forests in the Marathwada region has slightly increased from 3.48 per cent in 1981-82 to 3.51 per cent in 1999-2000. The percentage with respect to Maharashtra state has also slightly increased from 17.33 per cent in 1981-82 to 17.44 per cent in the year 1999-2000. Districtwise land use pattern reveals that the percentage of area under forests in Aurangabad district and Parbhani district declined from 7.64 per cent and 3.48 per cent in 1981-82 to 5.09 per cent and 3.18 per cent respectively in 1999-2000. But the percentage area under forests in Jalna,

Table 4
Districtwise Land Utilization Statistics in Marathwada Region

(Area in '00' hectares)

Sl. No.	Head of the classification	Aurangabad 1981-1982	Aurangabad 1999-2000	Jalna 1981-1982	Jalna 1999-2000	Beed 1981-1982	Beed 1999-2000	Latur 1981-1982	Latur 1999-2000
1.	Forests	770 (7.64)	513 (5.09)	52 (0.67)	56 (0.72)	215 (1.92)	233 (2.18)	8 (0.12)	18 (0.25)
2.	Barren & uncultivable land.	177 (1.76)	220 (1.98)	110 (1.42)	73 (0.94)	161 (1.44)	244 (2.28)	75 (1.12)	190 (2.65)
3.	Land put to non-agricultural uses	534 (5.30)	622 (6.17)	279 (3.61)	218 (2.82)	388 (3.42)	428 (4.00)	131 (1.96)	201 (2.81)
4.	Cultivable waste land	137 (1.36)	152 (1.24)	78 (1.00)	159 (2.06)	464 (4.15)	405 (3.79)	292 (4.38)	240 (3.35)
5.	Permanent pasture and other grazing land	445 (4.42)	375 (3.72)	294 (3.81)	262 (3.39)	416 (3.72)	382 (3.57)	188 (2.82)	210 (2.93)
6.	Land under miscellaneous trees, crops & groves not included in the net area sown	45 (0.45)	56 (0.56)	105 (1.36)	112 (1.45)	22 (0.20)	13 (0.12)	75 (1.12)	177 (2.47)
7.	Current fallows	117 (1.16)	389 (3.86)	253 (3.27)	726 (9.40)	1094 (9.79)	833 (7.80)	376 (5.63)	425 (5.94)
8.	Other fallows	117 (1.16)	542 (5.38)	234 (3.03)	160 (2.07)	593 (5.31)	396 (3.71)	182 (2.73)	370 (5.17)
9.	Net area sown	7629 (75.71)	7008 (69.54)	6321 (81.81)	5966 (77.22)	7816 (69.98)	7752 (72.54)	5347 (80.12)	5326 (74.42)

Contd...

(Area in '00' hectares)

Sl. No.	Head of the classification	Aurangabad 1981-1982	Aurangabad 1999-2000	Jalna 1981-1982	Jalna 1999-2000	Beed 1981-1982	Beed 1999-2000	Latur 1981-1982	Latur 1999-2000
10.	Area sown more than once	961	3299	631	1429	1092	1517	942	1828
		(9.54)	(32.74)	(8.17)	(18.50)	(9.78)	(14.20)	(14.11)	(25.54)
11.	Total cropped area	8590	10307	6952	7395	8908	9269	6289	7154
	(i.e. gross cropped area (9+10)	(85.24)	(102.28)	(89.98)	(95.72)	(79.76)	(86.74)	(94.23)	(99.96)
	Total geographical area	10077	10077	7726	7726	11169	10686	6674	7157
		(100)	(100)	(100)	(100)	(100)	(100)	(100)	(100)

Sl. No.	Head of the classification	Osmanabad 1981-1982	Osmanabad 1999-2000	Nanded 1981-1982	Nanded 1999-2000	Parbhani 1981-1982	Parbhani 1999-2000	Marathwada 1981-1982	Marathwada 1999-2000
1.	Forests	10	40	801	855	382	349	2238	2261
		(0.14)	(0.53)	(7.75)	(8.28)	(3.48)	(3.18)	(3.48)	(3.51)
2.	Barren & uncultivable land	79	66	205	187	139	169	946	1149
		(1.07)	(0.88)	(1.98)	(1.81)	(1.27)	(1.54)	(1.47)	(1.78)
3.	Land put to non-agricultural uses	133	155	402	358	375	390	2242	2372
		(1.81)	(2.07)	(3.89)	(3.47)	(3.42)	(3.55)	(3.49)	(3.68)
4.	Cultivable waste land	352	441	301	357	423	347	2047	2101
		(4.79)	(5.89)	(2.91)	(3.46)	(3.86)	(3.16)	(3.18)	(3.26)
5.	Permanent pasture and	176	185	662	595	383	310	2564	2319
	other grazing land	(2.39)	(2.47)	(6.41)	(5.76)	(3.49)	(2.83)	(3.99)	(3.60)

Contd...

(Area in '00' hectares)

Sl. No.	Head of the classification	Osmanabad 1981-1982	Osmanabad 1999-2000	Nanded 1981-1982	Nanded 1999-2000	Parbhani 1981-1982	Parbhani 1999-2000	Marathwada 1981-1982	Marathwada 1999-2000
6.	Land under miscellaneous trees, crops & groves not included in the net area sown	05 (0.07)	16 (0.21)	63 (0.61)	61 (0.59)	31 (0.28)	16 (0.15)	346 (0.54)	451 (0.70)
7.	Current fallows	409 (5.56)	883 (11.80)	283 (2.74)	694 (6.72)	288 (2.62)	786 (7.16)	2820 (4.39)	4736 (7.35)
8.	Other fallows	471 (6.41)	761 (10.17)	153 (1.48)	137 (1.33)	151 (1.38)	383 (3.49)	2007 (3.12)	2749 (2.27)
9.	Net area sown	5718 (77.76)	4938 (65.97)	7461 (72.22)	7087 (68.59)	8800 (80.20)	8222 (74.90)	49092 (76.35)	46299 (71.85)
10.	Area sown more than once	304 (4.14)	2469 (32.99)	378 (3.66)	1138 (11.02)	1159 (10.56)	4953 (45.10)	5467 (8.50)	16633 (25.82)
11.	Total cropped area (i.e. gross cropped area (9+10)	6022 (81.90)	7407 (98.96)	7839 (75.88)	8225 (79.61)	9959 (90.76)	13175 (120.0)	54559 (84.85)	62932 (97.67)
	Total geographical area	7353 (100)	7485 (100)	10331 (100)	10331 (100)	10972 (100)	10972 (100)	64302 (100)	64434 (100)

Sources: 1. Season & Crop Report of 1981-82 and 1999-2000.
2. Epitome of Agriculture of Maharashtra State, Part II, 2001-2002.

Beed, Latur, Osmanabad and Nanded districts has slightly increased during the period 1981-82 to 1999-2000.

The percentage of area under barren and uncultivable land in the Marathwada region has increased from 1.47 per cent in 1981-82 to 1.78 per cent in the year 1999-2000. Districtwise picture with respect to barren and uncultivable land shows that the percentage under it has slightly increased in Aurangabad, Beed, Latur and Parbhani districts but slightly declined in Jalna, Osmanabad and Nanded districts during the period under consideration.

The percentage of area under land put to non-agricultural uses in Marathwada region has slightly increased from 3.49 per cent in 1981-82 to 3.68 per cent in the year 1999-2000. The percentage of area under land put to non-agricultural uses has increased in Aurangabad, Beed, Latur, Osmanabad and Parbhani districts but slightly declined in Jalna and Nanded districts during the period under consideration.

The percentage of area under cultivable waste land in the Marathwada region has slightly increased from 3.18 per cent in the year 1981-82 to 3.26 per cent in the year 1999-2000. The percentage of area under cultivable waste land has increased in Jalna, Osmanabad and Nanded districts and declined in Aurangabad, Beed, Latur and Parbhani districts during this period.

The percentage of area under permanent pastures and other grazing land has slightly declined in Marathwada region from 3.99 per cent in the year 1981-82 to 3.60 per cent in the year 1999-2000. The percentage area under permanent pastures and other grazing land has slightly declined in Aurangabad, Jalna, Beed, Nanded and Parbhani districts but slightly increased in Latur and Osmanabad districts during the period from 1981-82 to 1999-2000.

The percentage of area under miscellaneous trees, crops and groves not included in net area sown has slightly increased in Marathwada region from 0.54 per cent in the year 1981-82 to 0.70 per cent in the year 1999-2000. The percentage of area under miscellaneous trees, crops and groves not included in net area sown has slightly increased in Aurangabad, Jalna,

Latur and Osmanabad districts and slightly decreased in Beed, Nanded and Parbhani districts during the period from 1981-82 to 1999-2000.

The percentage of area under current fallows in the Marathwada region has increased from 4.39 per cent in 1981-82 to 7.35 per cent in the year 1999-2000. The percentage of area under current fallows has increased in Aurangabad, Jalna, Latur, Osmanabad, Nanded and Parbhani districts but decreased in Beed district during the period from 1981-82 to 1999-2000.

The percentage of area under other fallows in the Marathwada region has decreased from 3.12 per cent in the year 1981-82 to 2.27 per cent in the year 1999-2000. The percentage of area under other fallows has increased in Aurangabad, Beed, Latur, Osmanabad and Parbhani districts but slightly decreased in Jalna and Nanded districts during the period from 1981-82 to 1999-2000.

The percentage of net area sown in the Marathwada region has decreased from 76.35 per cent in the year 1981-82 to 71.85 per cent in the year 1999-2000. The percentage of net area sown has decreased in Aurangabad, Jalna, Latur, Osmanabad, Nanded and Parbhani districts but increased in only Beed district during the period from 1981-82 to 1999-2000.

The percentage of area sown more than once in Marathwada region has significantly increased from 8.50 per cent in the year 1981-82 to 25.82 per cent in the year 1999-2000 as well as the percentage of area sown more than once has also increased in all the districts of Marathwada region during the period from 1981-82 to 1999-2000.

Lastly, the percentage of gross cropped area in Marathwada region has significantly increased from 84.85 per cent in the year 1981-82 to 97.67 per cent in the year 1999-2000. As well as the percentage of gross cropped area in all the districts of Marathwada region has also increased during the period from 1981-82 to 1999-2000.

IRRIGATION

Irrigation plays an important role in the agricultural economy of the region in raising agricultural production and productivity. It is one of the important factors for assured production of various crops in the region. Irrigation leads to better utilization of other factors of production and leads to increase in yield per hectare of land. Irrigation facility creates employment opportunities in the region which is important in the process of economic development. Irrigation, therefore, is useful to achieve self-sufficiency in food production by producing higher level of agricultural production.

The following Table 5 shows the percentage of gross area irrigated to gross cropped area in all the districts of Marathwada region. This percentage has continuously increased from 10.59 per cent in 1981-82 to 16.80 per cent in Marathwada in the year 1998-99. In the Maharashtra state it has increased from 12.46 per cent in 1981-82 to 17.90 in the year 1998-99. The percentage of gross area irrigated to gross cropped area has continuously increased in all the districts of Marathwada region during the period from 1981-82 to 1998-99. It has increased in Aurangabad, Jalna, Parbhani, Beed, Nanded, Osmanabad and Latur districts from 14.71 per cent11.21 per cent,7.69 per cent,13.77 per cent, 5.92 per cent, 14.15 per cent and 6.74 per cent in 1981-82 to 21.9 per cent, 14.1 per cent, 11.7 per cent, 27 per cent, 10.9 per cent, 17.6 per cent and 7.7 per cent in 1998-99 respectively.

The percentage of gross area irrigated to gross cropped area is the highest (27 per cent) in Beed district and lowest (7.7 per cent) in Latur district in the year 1998-99. In Beed, Aurangabad and Osmanabad districts the percentage of gross irrigated area is the highest than the percentage of gross irrigated area for the region as a whole. In Jalna, Parbhani, Nanded and Latur districts the percentage of gross irrigated area is less than the percentage of gross irrigated area of the region as a whole for the period 1998-99.

Table 5 shows that the well is the main source of irrigation in all the districts of Marathwada region during the period from 1981-82 to 1998-99. The percentage of area irrigated by wells to total net area irrigated in Marathwada has increased from 71.18 per cent in 1981-82 to 75.07 per cent in the year 1998-99. The percentage of area irrigated by wells is the highest

Table 5
Area Irrigated by Different Sources in Marathwada Region

(Area in '00' hectares)

Sl. No.	Source of Irrigation	Aurangabad 1981-1982	Aurangabad 1998-1999	Jalna 1981-1982	Jalna 1998-1999	Parbhani 1981-1982	Parbhani 1998-1999	Beed 1981-1982	Beed 1998-1999
1.	Surface (Net)	163	232	109	88	299	397	237	711
		(21.42)	(14.26)	(20.34)	(12.83)	(53.01)	(42.14)	(22.11)	(33.13)
2.	Wells (Net)	598	1395	427	598	265	545	835	1435
		(78.58)	(85.74)	(79.66)	(71.17)	(46.99)	(57.86)	(77.89)	(66.87)
3.	Total net area irrigated	761	1627	536	686	564	942	1072	2146
		(100)	(100)	(100)	(100)	(100)	(100)	(100)	(100)
4.	Total (gross) area irrigated	1264	1872	779	1026	766	1374	1227	2536
5.	Total (Gross) cropped area	8590	8548	6952	7277	9959	11744	8908	9393
6.	Percentage of total (gross) area irrigated to total (gross) cropped area	(14.71)	(21.9)	(11.21)	(14.1)	(7.69)	(11.7)	(13.77)	27

Contd...

Sl. No.	Source of Irrigation	Nanded 1981-1982	Nanded 1998-1999	Osmanabad 1981-1982	Osmanabad 1998-1999	Latur 1981-1982	Latur 1998-1999	Marathwada 1981-1982	Marathwada 1998-1999
1.	Surface (Net)	221	227	187	207	42	42	1258	1904
		(59.25)	(33.78)	(26.19)	(19.13)	(12.17)	(8.73)	(28.82)	(24.93)
2.	Wells (Net)	152	445	527	875	303	439	3107	5732
		(40.75)	(66.22)	(73.81)	(80.87)	(87.83)	(91.27)	(71.18)	(75.07)
3.	Total net area irrigated	373	672	714	1082	345	481	4365	7636
		(100)	(100)	(100)	(100)	(100)	(100)	(100)	(100)
4.	Total (gross) area irrigated	464	902	852	1339	424	534	5776	9583
5.	Total (Gross) cropped area	7839	8276	6022	7608	6289	6936	54559	57042
6.	Percentage of total (gross) area irrigated to total (gross) cropped area	5.92	10.9	14.15	17.6	6.74	7.7	10.59	16.80

Source: 1. Season & Crop Report of 1981-82 & 1998-99.
2. Epitome of Agriculture in Maharashtra, Part II, 2001-02.

i.e. 91.27 per cent in Latur district and the lowest i.e. 57.86 per cent in Parbhani district in the year 1998-99. The percentage of area irrigated by well has increased in Aurangabad, Parbhani, Nanded, Osmanabad and Latur districts from 78.58 per cent, 46.99 per cent, 40.75 per cent, 73.81 per cent and 87.83 per cent in the year 1981-82 to 85.74 per cent, 57.86 per cent, 66.22 per cent, 80.87 per cent and 91.27 per cent in the year 1998-1999 respectively. The percentage of well irrigated area is declined in Jalna and Beed districts from 79.66 per cent and 77.89 per cent in the year 1981-82 to 71.17 per cent and 66.87 per cent in the year 1998-99 respectively. In Jalna, Parbhani, Beed and Nanded districts the percentage of area irrigated by well is less than the percentage of area irrigated by well for the region as a whole in 1998-99. And in Aurangabad, Osmanabad and Latur districts the percentage of area irrigated by well is higher than the percentage of area irrigated by well for the region as a whole, in the year 1998-99.

The percentage of area irrigated under surface irrigation (surface irrigation including area irrigated by canals, tanks and other sources of irrigation) in Marathwada region has decreased from 28.82 per cent in the year 1981-82 to 24.93 per cent in the year 1998-1999. The districtwise percentage area irrigated by surface irrigation is 14.26 per cent, 12.83 per cent, 42.14 per cent, 33.13 per cent, 33.78 per cent, 19.13 per cent and 8.73 per cent in Aurangabad, Jalna, Parbhani, Beed, Nanded, Osmanabad and Latur districts, respectively in the year 1998-1999. The percentage area under surface irrigation is the highest (i.e. 59.25 per cent) in Nanded district and the lowest (i.e. 12.17 per cent) in Latur district in the year 1981-82. It is the highest (i.e. 42.4 per cent) in Parbhani district and the lowest (i.e. 8.73 per cent) in Latur district in the year 1998-99. It is clear that the percentage of area under surface irrigation is highest in Parbhani district in the year 1981-82 and 1998-99 also. As well as the percentage of area under surface irrigation is lowest in Latur district in the year 1981-82 and 1998-99 also. Only in Beed district the percentage of area under surface irrigation is increased over the period under consideration. In 1998-99, 75.07 per cent net area irrigated was through wells only.

POPULATION

Districtwise population and decadal variation rates in Marathwada region are given in Table 6. From this table it is clear that the population of the Marathwada region as per 2001 census is 155.89 lakhs. It was 62.97 lakh in the year 1961. There was a rising trend in the population from the decade 1951-61 to 1961-71. However, it was increased at a decreasing rate (20.73 per cent) in the decade 1971-81. But increased at an increasing rate of 31.56 per cent during the decade 1981-91. During the last decade (1991-2001) the population has increased at a decreasing rate of 21.78 per cent.

As per the 2001 census, the population is the highest (26.68 lakh) in Aurangabad district and lowest (21.59 lakh) in Beed district. During the last decade (1991-2001) the population of Aurangabad district has increased at decreasing rate (26.68 per cent) which is the highest rate in the Marathwada region. The percentage of last decade variation in the population was the lowest (17.04 per cent) in Parbhani district. The total population of Marathwada region has increased from 51.09 lakh in the year 1951 to 155.89 lakh in the year 2001. It means that the population has increased three fold during the 50 years period. But in the last decade population has increased at decreasing rate (21.78 per cent) in the Marathwada region.

RURAL AND URBAN POPULATION

Table 7 shows the percentage of rural and urban population in Marathwada region in the year 2001. It is clear that 75.41 per cent population lives in rural area and 24.59 per cent population lives in urban area of Marathwada region. Table 7 shows that the percentage of rural population is highest (84.40 per cent) in Hingoli district and lowest (62.82 per cent) in Aurangabad district. The percentage of urban population is highest (37.18 per cent) in Aurangabad district and lowest (15.60 per cent) in Hingoli district in the year 2001. The percentage of rural population in Jalna, Hingoli, Beed, Nanded, Osmanabad and Latur districts is higher than the percentage of rural population of Marathwada region as a whole in the year 2001.

Table 6
Districtwise Population and Decadal Variation Rates in Marathwada

	Aurangabad		*Beed*		*Parbhani*		*Nanded*	
Year	*Population*	*Decade variation (in %)*	*Population*	*Decade variation (in %)*	*Population*	*Decade variation (in %)*	*Population*	*Decade variation (in %)*
1951	11,79,404	+ 8.93	8,26,046	+13.44	10,10,864	+10.54	8,83,531	+12.65
1961	15,32,341	+29.93	10,01,466	+21.24	12,06,236	+19.33	10,79,674	+22.20
1971	19,71,006	+28.63	12,86,121	+28.42	15,06,771	+24.92	13,97,762	+29.46
1981	24,33,420	+23.46	14,86,030	+15.54	18,29,338	+21.41	17,49,334	+25.15
1991	35,78,244	+47.08	18,22,072	+22.61	21,17,035	+15.72	23,30,374	+33.19
2001	45,32,905	+26.68	21,59,841	+18.54	24,77,826	+17.04	28,68,158	+23.08

Contd...

	Osmanabad		Jalna		Latur		Marathwada	
Year	Population	Decade variation (in %)	Population	Decade variation (in %)	Population	Decade variation (in %)	Population	Decade variation (in %)
1951	12,10,041	+ 9.85	-	-	-	-	51,09,886	+10.82
1961	14,77,656	+22.12	-	-	-	-	62,97,373	+23.24
1971	18,96,687	+28.36	-	-	-	-	80,58,347	+27.96
1981	22,30,620	+17.61	-	-	-	-	97,28,782	+20.73
1991	29,52,968	+32.39	13,64,465	-	16,76,641	-	1,28,00,653	+31.56
2001	35,50,493	+20.23	16,12,357	18.17	20,78,237	23.95	1,55,89,223	+21.78

Notes: 1. Population of Jalna district is included in Aurangabad district.
2. Population of Latur district is included in Osmanabad district.
3. Population of Hingoli district is included in Parbhani district.

Sources: 1. Census of India, District Census Handbook of All Districts in Marathwada Region.
2. Census of Maharashtra, 1981, Series 12 and 1991, Series 14 and 2001.
3. Maharashtra, 2003.
4. Human Development Report, Government of Maharashtra, 2002.

Table 7
Districtwise Rural and Urban
Population in Marathwada Region as per Census of 2001

(Figures in '000')

	2001		
District	*Rural*	*Urban*	*Total*
Aurangabad	1835	1086	2921
	(62.82)	(37.18)	(100)
Jalna	1304	308	1612
	(80.89)	(19.11)	(100)
Parbhani	1006	485	1491
	(67.47)	(32.53)	(100)
Hingoli	833	154	987
	(84.40)	(15.60)	(100)
Beed	1773	387	2160
	(82.08)	(17.92)	(100)
Nanded	2173	689	2862
	(75.93)	(24.07)	(100)
Osmanabad	1239	233	1472
	(84.17)	(15.83)	(100)
Latur	1588	490	2078
	(76.42)	(23.58)	(100)
Marathwada	11751	3832	15583
	(75.41)	(24.59)	(100)

Note: Figures in brackets indicate percentage to total population.

Sources: 1. Population Census 2001.
2. Human Development Report, Government of Maharashtra, 2002.

POPULATION OF MALES AND FEMALES

Districtwise population of males and females in Marathwada region as per the census of 2001 is given in the Table 8. The percentage of male population is 51.63 and female population percentage is 48.37 in Marathwada region. It is clear that the percentage of male population is higher than the female population in the region by 3.26 per cent.

Table 8 indicates that the percentage of male population is highest (52.10 per cent) in Aurangabad district and lowest (51.10 per cent) in Parbhani district in the year 2001. The percentage of female population is highest (48.90 per cent) in Parbhani district and lowest (47.90 per cent) in Aurangabad district in the year 2001. The percentage of male population in all the districts of Marathwada region is higher than the percentage of female population in all the districts of the region.

Table 8
Districtwise Population of Males and Females in Marathwada Region as per the Census of 2001

	2001		
District	*Male*	*Female*	*Total*
Aurangabad	1521632	1398916	2920548
	(52.10)	(47.90)	(100)
Jalna	825977	786380	1612357
	(51.23)	(48.77)	(100)
Parbhani	761937	729172	1491109
	(51.10)	(48.90)	(100)
Hingoli	505188	481529	986717
	(51.20)	(48.80)	(100)
Beed	1120664	1039177	2159841
	(51.89)	(48.11)	(100)
Nanded	1476300	139185	2868158
	(51.47)	(48.53)	(100)
Osmanabad	762947	709309	1472256
	(51.82)	(48.18)	(100)
Latur	1074321	1003916	2078237
	(51.69)	(48.31)	(100)
Marathwada	8048959	7540256	15589223
	(51.63)	(48.37)	(100)

Note: Figures in the brackets indicate percentage of population.
Sources: 1. Population census, 2001.
2. Human Development Report, Government of Maharashtra, 2002.

Table 8 shows that the percentage of male population in Aurangabad district is 52.10 per cent which is higher than other districts and the percentage of female population is relatively higher in Parbhani district i.e. 48.90 per cent.

DISTRICTWISE WORKERS

Table 9 shows the districtwise number of workers in Marathwada region as per the census of 2001. In Marathwada region, total workers are 58,28,469 which are 37.39 per cent of

Table 9

Districtwise Number of Workers in Marathwada Region as per the Census of 2001 (Number)

District	Rural			Urban			Total		
	Male	*Female*	*Total*	*Male*	*Female*	*Total*	*Male*	*Female*	*Total*
Aurangabad	449208	324752	773960	250222	44506	294728	699430	369258	1068688
	(15.38)	(11.12)	(26.50)	(8.57)	(1.52)	(10.09)	(23.95)	(12.64)	(36.60)
Jalna	316305	234172	550477	68052	10688	78740	384410	244807	629217
	(19.62)	(14.52)	(34.14)	(4.22)	(0.66)	(4.88)	(23.84)	(15.18)	(39.02)
Parbhani	259641	191316	450957	95678	18122	113800	355319	209438	564757
	(17.41)	(12.83)	(30.24)	(6.42)	(1.22)	(7.63)	(23.83)	(14.05)	(37.87)
Hingoli	209995	172661	382656	33319	6712	40031	243314	179373	422687
	(21.28)	(17.50)	(38.78)	(3.38)	(0.68)	(4.06)	(24.66)	(18.18)	(42.84)*
Beed	423215	334755	757970	82106	13486	95592	505321	348241	853562
	(19.59)	(15.50)	(35.09)	(3.80)	(0.62)	(4.43)	(23.40)	(16.12)	(39.52)
Nanded	496326	326647	822973	144934	24344	169278	641266	350991	992251
	(17.30)	(11.39)	(28.69)	(5.05)	(0.85)	(5.90)	(22.36)	(12.24)	(34.60)*
Osmanabad	305415	200026	505441	48773	9470	58243	354188	209496	563684
	(20.74)	(13.59)	(34.33)	(3.31)	(0.64)	(3.96)	(24.06)	(14.23)	(38.29)
Latur	387463	225246	612709	105859	15061	120920	493322	240307	733629
	(18.64)	(10.84)	(29.48)	(5.09)	(0.72)	(5.82)	(23.74)	(11.56)	(35.30)
Marathwada	2847559	2009578	4857137	828943	142389	1971332	3676555	2151914	5828469
	(18.27)	(12.89)	(31.16)	(5.32)	(0.91)	(6.23)	(23.59)	(13.80)	(37.39)

Sources: 1. Human Development Report, 2002, Government of Maharashtra.
2. Population Census, 2001.

the total population of the region. Out of them 23.59 per cent are male workers and 13.80 per cent are female workers. Population of male workers is higher than the population of female workers in the region by 9.79 percentage points. The percentage of rural workers is 31.16 and percentage of urban workers is only 6.23. Table 9 indicates that the population of rural workers is higher than the population of urban workers by 24.93 percentage points in Marathwada region. In Marathwada region the percentage of rural male workers is 18.27 and rural female workers percentage is 12.84. It means the percentage of rural male workers is higher than the percentage of rural female workers by 5.38 percentage points in the region. The percentage of urban male workers is 5.32 and the percentage of urban female workers is 0.91. Here, the percentage of rural female workers is very negligible. The percentage of rural male workers (18.27) is higher than the percentage of urban male workers (5.32) in the region by 12.95 percentage points. As well as the percentage of rural female workers (21.89) is higher than the percentage of urban female workers (0.91) by 11.98 percentage points in Marathwada region.

Table 9 shows that the percentage (i.e. 14.84 per cent) of total workers to the total population is highest in Hingoli district and lowest (i.e. 34.60 per cent) in Nanded district. The percentage (i.e. 38.78 per cent) of total rural workers is highest in Hingoli district and lowest (i.e. 26.50 per cent) in Aurangabad district. The percentage (i.e. 10.09 per cent) of total urban workers is highest in Aurangabad district and lowest (i.e. 3.96 per cent) in Osmanabad district. The percentage (i.e. 21.28 per cent) of rural male workers is highest in Hingoli district and lowest (i.e. 15.38 per cent) in Aurangabad district. The percentage (i.e. 17.50 per cent) of rural female workers is higher in Hingoli district and lowest (i.e. 10.84 per cent) in Latur district. The percentage (i.e. 8.57 per cent) of urban male workers is highest in Aurangabad district and lowest (i.e. 3.31 per cent) in Osmanabad district. The percentage (i.e. 1.52 per cent) of urban female workers is highest in Aurangabad district and lowest (i.e. 0.62 per cent) in Beed district.

However, the percentage of total workers to the total population in Jalna, Parbhani, Hingoli, Beed and Osmanabad districts is higher than the per centage of total workers to the total population in Marathwada region as a whole and lower in Aurangabad, Nanded and Latur districts in the year 2001.

DISTRIBUTION OF WORKERS BY ACTIVITY

The distribution of workers by economic activity in 2001 in the districts of Marathwada region is given in Table 10. From this table it is clear that agriculture is an important occupation in Marathwada region and involves 77 per cent of earners (i.e. farmers and agricultural labourers) in the region. The

Table 10
Districtwise Distribution of Workers in Marathwada Region by Economic Activity During the Year 2001

District	*Economic Activity* Farmers	Agri-cultural Labourers	Forestry, Animal Husbandry & Fishery	Trade & Commerce	Other	Total
Aurangabad	365 (37.05)	251 (25.48)	110 (11.17)	51 (5.18)	208 (21.12)	985 (100)
Jalna	281 (46.83)	214 (35.67)	7 (1.17)	26 (4.33)	72 (12.00)	600 (100)
Parbhani	347 (38.99)	369 (41.46)	12 (1.35)	40 (4.49)	12.2 (13.71)	890 (100)
Beed	373 (49.08)	231 (30.39)	13 (1.71)	33 (4.34)	110 (14.48)	760 (100)
Nanded	341 (36.78)	383 (41.32)	12 (1.29)	48 (5.18)	143 (15.43)	927 (100)
Osmanabad	215 (40.65)	217 (41.02)	6 (1.13)	18 (3.40)	73 (13.80)	529 (100)
Latur	258 (39.33)	257 (39.18)	5 (0.76)	36 (5.49)	100 (15.24)	656 (100)
Marathwada	2180 (40.77)	1922 (35.95)	165 (3.09)	252 (4.71)	828 (15.48)	5347 (100)

Note: 1. Figures for newly created Hingoli District are included in undivided Parbhani District.
2. Figures in the brackets indicate percentage of workers.

Sources: At A Glance-Aurangabad Division, Office of Economics and Statistics, Aurangabad, 2002.

percentage of farmers to the total workers in region is 40.77 per cent and percentage of agricultural labourers to the total workers in region is 35.95 per cent. The percentage of other workers (i.e. forestry, animal husbandry, fishery, trade and commerce) is 23 per cent to the total workers.

The percentage of farmers to total workers is comparatively higher in Beed district (i.e. 49.08 per cent) and Jalna district (i.e. 46.83 per cent). The percentage of farmers in Aurangabad, Parbhani, Nanded, Osmanabad and Latur districts is 37.05 per cent, 38.99 per cent, 36.78 per cent, 40.65 per cent and 39.33 per cent respectively in the year 2001. Another feature of this distribution of workers is that the large size of the class of agricultural labourers. The percentage of agricultural labourers to the total workers is very high in the region because of the lack of alternative job opportunities in the non-agricultural sector of the region. Within the region there are considerable variations among different districts with respect to the proportion of agricultural labourers. The proportion of agricultural labour is noted to be particularly high in Parbhani, Nanded, Osmanabad and Latur districts of the region.

Table 11 shows the district wise land holders in Marathwada region during the year 1991-92. From this table it is clear that the percentage of 2 to 10 hectare land holders (i.e. 41.12 per cent) is high in the region and more than 10 hectare land holders to total land holders is very low i.e. 1.42 per cent. The percentage of less than 1 hectare land holders is 24.21 and a 1 to 2 hectare land holder is 33.25 per cent in the region.

Table 11 indicates that the percentage of less than 1 hectare land holders is very high (i.e. 30.45 per cent) in Beed district and low (i.e. 17.20 per cent) in Latur district. The percentage of 1 to 2 hectare land holders is high (i.e. 35.59 per cent) in Jalna district and low (i.e. 30.09 per cent) in Osmanabad district. The percentage of 2 to 10 hectare land holders is high (i.e. 48.96 per cent) in Osmanabad district and low (i.e. 36.49 per cent) in Beed district. And the percentage of more than 10 hectare land holders is high (i.e. 3.23 per cent) in Osmanabad district and very low (i.e. 0.68 per cent) in Nanded district in the year 1991-92.

Table 11
Districtwise Land Holders in Marathwada Region During the Year 1991-92

(Figures in '00')

District	*Less than 1 hectare*	*1 to 2 hectares*	*2 to 10 hectares*	*More than 10 hectares*	*Total*
Aurangabad	828	1177	1344	42	3391
	(24.42)	(34.71)	(39.63)	(1.24)	(100)
Jalna	672	984	1077	32	2765
	(24.30)	(35.59)	(38.95)	(1.16)	(100)
Parbhani	900	1264	1702	43	3909
	(23.02)	(32.34)	(43.54)	(1.10)	(100)
Beed	1235	1287	1480	54	4056
	(30.45)	(31.73)	(36.49)	(1.33)	(100)
Nanded	1005	1330	1434	26	3795
	(26.48)	(35.05)	(37.79)	(0.68)	(100)
Osmanabad	384	652	1061	70	2167
	(17.72)	(30.09)	(48.96)	(3.23)	(100)
Latur	399	756	1114	51	2320
	(17.20)	(32.58)	(48.02)	(2.20)	(100)
Marathwada	5423	7450	9211	319	22403
	(24.21)	(33.25)	(41.12)	(1.42)	(100)

Note: 1. Figures in brackets indicate percentage of landholders.

Sources: 1. At A Glance-Aurangabad Division, Office of Economics and Statistics, Aurangabad, 2002.
2. Directorate of Economics & Statistics, Govt. of Maharashtra.

CROPPING PATTERN

The cropping pattern indicates the allocation of land to different crops by the farmers. The agricultural development of a region depends upon the cropping pattern of the region. The cropping pattern affects the general employment and total earning of the people. The major crops grown in the Marathwada region are the kharif jowar, rabi jowar, bajra, tur, gram, cotton, sugarcane, wheat, groundnut, safflower, sunflower and soyabean.

Cropping pattern in Marathwada region is shown in Table 12. It shows that the gross cropped area in the Marathwada region has increased from 54,559 hundred hectares in 1981-82 to 59,893 hundred hectares in 1998-99. It means the gross cropped area has increased by 5,334 hundred hectares. The bulk of the increase is due to the extension of the net cultivable area in the region.

Table 12
Cropping Pattern in Marathwada Region

(Area in '00' hectares)

	Districts							
	Aurangabad		*Jalna*		*Beed*		*Parbhani*	
Crop	*1981-1982*	*1998-1999*	*1981-1982*	*1998-1999*	*1981-1982*	*1998-1999*	*1981-1982*	*1998-1999*
Rice	51	13	37	11	67	35	228	229
	(0.59)	(0.15)	(0.53)	(0.15)	(0.75)	(0.37)	(2.29)	(1.94)
Wheat	526	515	410	305	466	407	565	739
	(6.12)	(6.04)	(5.90)	(4.18)	(5.23)	(4.33)	(5.67)	(6.27)
Kharif	610	294	702	360	1367	734	1977	1844
jowar	(7.10)	(3.45)	(10.10)	(4.93)	(15.35)	(7.82)	(19.85)	(15.64)
Rabi	2537	1827	1442	1305	2687	2391	1730	1491
jowar	(29.53)	(21.42)	(20.74)	(17.88)	(30.16)	(25.46)	(17.37)	(12.64)
Total	3147	2121	2144	1665	4054	3125	3707	3535
jowar	(36.64)	(24.87)	(30.84)	(22.81)	(45.51)	(33.28)	(37.22)	(28.28)
Bajra	1463	1330	409	800	1306	1768	88	94
	(17.103)	(15.59)	(5.88)	(10.96)	(14.66)	(0.19)	(0.88)	(0.80)
Other	5	1	7	4	00	21	41	9
cereals	(0.06)	(0.01)	(0.10)	(0.05)		(0.22)	(0.41)	(0.08)
Total	5247	4437	3036	3198	5559	5409	4644	4426
cereals	(61.08)	(52.05)	(43.67)	(43.81)	(66.89)	(57.60)	(46.63)	(37.53)
Tur	315	409	259	466	348	461	396	732
	(3.67)	(4.80)	(3.73)	(6.38)	(3.91)	(4.91)	(3.98)	(6.21)
Gram	261	485	131	157	378	236	302	6.81
	(3.04)	(5.69)	(1.88)	(2.15)	(4.24)	(2.51)	(3.03)	(5.77)
Other	00	00	28	92	00	18	48	7
pulses			(0.40)	(1.26)		(0.19)	(0.48)	(0.06)
Total	1476	1239	1537	1551	1289	997	1872	2522
pulses	(17.18)	(14.53)	(22.11)	(21.25)	(14.47)	(10.62)	(18.80)	(21.39)
Total	6723	5676	4573	4749	7248	6406	6516	6948
Foodgrains	(78.27)	(66.55)	(65.78)	(65.05)	(81.37)	(68.21)	(65.43)	(58.92)
Sugarcane	220	185	85	87	67	204	52	180
	(2.56)	(2.17)	(1.22)	(1.19)	(0.75)	(2.17)	(0.52)	(1.53)
Cotton	581	1555	1302	1620	350	1110	2100	3300
	(6.76)	(18.23)	(18.73)	(22.19)	(3.93)	(11.82)	(21.09)	(27.98)

Contd...

(Area in '00' hectares)

	Districts							
	Aurangabad		*Jalna*		*Beed*		*Parbhani*	
Crop	*1981-1982*	*1998-1999*	*1981-1982*	*1998-1999*	*1981-1982*	*1998-1999*	*1981-1982*	*1998-1999*
Groundnut	166 (1.93)	136 (1.59)	87 (1.25)	37 (0.51)	148 (1.66)	153 (1.63)	150 (1.51)	125 (1.06)
Total oilseeds	921 (10.72)	920 (10.79)	875 (12.59)	721 (9.88)	1113 (12.49)	1095 (11.66)	1044 (10.48)	1151 (9.76)
Safflower	587 (6.83)	343 (4.02)	630 (9.06)	331 (4.53)	463 (5.20)	258 (2.75)	555 (5.57)	482 (4.09)
Sunflower	3 (0.03)	237 (2.78)	5 (0.07)	200 (2.74)	59 (0.66)	474 (5.05)	4 (0.04)	320 (2.71)
Soyabean	-	37 (0.43)	-	45 (0.62)	-	11 (0.12)	2.00 (0.02)	90 (0.76)
Gross cropped area	8590 (100)	8529 (100)	6952 (100)	7300 (100)	8908 (100)	9391 (100)	9959 (100)	11793 (100)

(Area in '00' hectares)

	Districts							
	Nanded		*Osmanabad*		*Latur*		*Marathwada*	
Crop	*1981-1982*	*1998-1999*	*1981-1982*	*1998-1999*	*1981-1982*	*1998-1999*	*1981-1982*	*1998-1999*
Rice	355 (4.53)	253 (3.05)	129 (2.14)	148 (1.94)	197 (3.13)	202 (2.90)	1064 (1.95)	891 (1.49)
Wheat	400 (5.10)	295 (3.56)	334 (5.55)	422 (5.53)	254 (4.04)	277 (3.98)	2955 (5.42)	2960 (4.94)
Kharif jowar	2528 (32.25)	2017 (24.33)	1042 (17.30)	985 (12.92)	1974 (31.39)	1760 (25.29)	10200 (18.70)	7994 (13.34)
Rabi jowar	497 (6.34)	266 (3.21)	1812 (30.09)	1867 (24.48)	350 (5.57)	236 (3.38)	11055 (20.26)	9383 (15.67)
Total jowar	3025 (38.59)	2283 (27.54)	2854 (47.39)	2852 (37.40)	2324 (36.95)	1996 (28.67)	21255 (38.96)	17377 (29.01)
Bajra	21 (0.27)	3 (0.04)	34 (0.56)	176 (2.31)	125 (1.99)	101 (1.45)	3446 (6.31)	4272 (7.13)
Other cereals	00	5 (0.06)	16 (0.27)	15 (0.20)	16 (0.25)	10 (0.14)	85 (0.16)	65 (0.11)
Total cereals	3824 (48.78)	2852 (34.40)	3467 (57.57)	3711 (48.66)	2992 (47.58)	3638 (52.25)	29169 (53.46)	27671 (46.20)

Contd...

(Area in '00' hectares)

	Districts							
	Nanded		*Osmanabad*		*Latur*		*Marathwada*	
Crop	*1981-1982*	*1998-1999*	*1981-1982*	*1998-1999*	*1981-1982*	*1998-1999*	*1981-1982*	*1998-1999*
Tur	370	481	534	749	505	667	2727	3965
	(4.72)	(5.81)	(8.87)	(9.82)	(8.03)	(9.58)	(5.00)	(6.62)
Gram	154	324	349	500	257	289	1832	2672
	(1.96)	(3.91)	(5.80)	(6.56)	(4.09)	(4.15)	(3.365)	(4.46)
Other pulses	00	00	43	00	342	16	461	133
			(0.71)		(5.44)	(0.23)	(0.84)	(0.22)
Total pulses	1115	1648	1401	1744	1659	2033	10349	11734
	(14.22)	(19.88)	(23.26)	(22.87)	(26.38)	(29.15)	(18.97)	(19.59)
Total	4939	4500	4868	5455	4651	4671	39518	38405
Foodgrains	(63.00)	(54.28)	(80.84)	(71.53)	(73.95)	(67.08)	(72.43)	(64.12)
Sugarcane	71	135	126	209	72	245	693	1245
	(0.91)	(1.63)	(2.09)	(2.74)	(1.44)	(3.52)	(1.27)	(2.08)
Cotton	2128	2781	1	40	289	316	6751	10722
	(27.15)	(33.54)	(0.02)	(0.52)	(4.60)	(4.54)	(12.37)	(17.90)
Groundnut	184	112	178	145	303	193	1216	901
	(2.35)	(1.35)	(2.96)	(1.90)	(4.82)	(2.77)	(2.23)	(1.50)
Total	395	634	876	1612	1032	1521	6256	7654
oilseeds	(5.04)	(7.65)	(14.55)	(21.14)	(16.41)	(21.84)	(11.47)	(12.78)
Safflower	21	79	308	370	124	191	2688	2054
	(0.27)	(0.95)	(5.11)	(4.85)	(1.97)	(2.74)	(4.93)	(3.43)
Sunflower	4	251	98	794	206	795	379	3071
	(0.05)	(3.03)	(1.63)	(10.41)	(3.28)	(11.42)	(0.69)	(5.13)
Soyabean	-	61	1	19	1	49	4	312
		(0.74)	(0.02)	(0.25)	(0.02)	(0.70)	(0.01)	(0.52)
Gross	7839	8291	6022	7626	6289	6963	54559	59893
cropped area	(100)	(100)	(100)	(100)	(100)	(100)	(100)	(100)

Source: Season and Crop Reports of the Year 1981-82 and 1998-99.

The cropping pattern in the Marathwada region is dominated by foodgrain crops. The area under foodgrains has occupied about 72.43 per cent in the year 1981-82. It has declined to 64.12 per cent in the year 1998-99.The percentage of area under total cereals in Marathwada region has decreased from 53.46 per cent in 1981-82 to 46.20 per cent in 1998-99. The percentage of area under tur has slightly increased from 5.00

per cent in 1981-82 to 6.62 per cent in 1998-99. The percentage of area under gram also slightly increased from 3.36 per cent in 1981-82 to 4.46 per cent in 1998-99. The percentage of area under total pulses has slightly increased from 18.97 per cent in 1981-82 to 19.59 per cent in 1998-99.

The percentage of area under sugarcane in Marathwada region had occupied about 1.27 per cent in 1981-82 which has increased to 2.08 per cent in 1998-99. The percentage of area under cotton in Marathwada region has significantly increased from 12.37 per cent in 1981-82 to 17.90 per cent in the year 1998-99. The percentage of area under groundnut has slightly declined from 2.23 per cent in 1981-82 to 1.50 per cent in 1998-99. The percentage of area under safflower has slightly declined from 4.93 per cent in 1981-82 to 3.43 per cent in 1998-99. The percentage of area under sunflower has significantly increased from 0.69 per cent in 1981-82 to 5.13 per cent in 1998-99. The percentage of area under soyabean in Marathwada region has occupied only 0.01 per cent in the year 1981-82 which has slightly increased to 0.52 per cent in 1998-99. The percentage of area under total oilseeds in the Marathwada region has slightly increased from 11.47 per cent in 1981-82 to 12.78 per cent in the year 1998-99.

From the above analysis of cropping pattern it is clear that the percentage of area under foodgrain crops as compared to other crops in Marathwada region is the highest and showed declining trend during the period under consideration. Among the cash crops, the percentage of area under sugarcane and cotton has increased. The percentage of area under total oilseeds has increased but percentage of area under groundnut and safflower has declined in the Marathwada region during the period under consideration.

Table 13 shows the districtwise gross cropped area and gross irrigated area in Marathwada region during the period 1981-82 and 1999-2000. From the Table 13 it is clear that the percentage of gross irrigated area to gross cropped area in Marathwada region has increased from 10.59 per cent in 1981-82 to 14.33 per cent in the year 1999-2000. The percentage of gross irrigated area to gross cropped area has substantially increased in all the districts of Marathwada region except Latur district.

Table 13
Districtwise Gross Cropped and Gross Irrigated Area in Marathwada Region

Districts	*Year*	*Gross cropped area (GCA)*	*Gross irrigated area (GIA)*	*Percentage of GIA to GCA*
Aurangabad	*1981-82*	*8590*	*1264*	14.71
	1999-00	10307	1900	18.43
Jalna	1981-82	6592	779	11.21
	1999-00	7395	1110	15.01
Beed	1981-82	8908	1227	13.77
	1999-00	9269	2530	27.30
Osmanabad	1981-82	6022	852	14.15
	1999-00	7407	1220	16.47
Latur	1981-82	6289	424	6.74
	1999-00	7154	390	5.45
Parbhani	1981-82	9959	766	7.69
	1999-00	13175	1120	8.50
Nanded	1981-82	7839	464	5.92
	1999-00	8225	750	9.12
Marathwada	1981-82	54,559	5776	10.59
	1999-00	62,939	9020	14.33

Sources: 1. Season and crop Report, Maharashtra State, 1981-82 and 1999-2000.
2. Office of the Commissionerate of Agriculture, Maharashtra State Pune.

The percentage of gross irrigated area to gross cropped area is highest (i.e. 27.30 per cent) in Beed district and the lowest (i.e. 5.45 per cent) in Latur district in the year 1999-2000.

CONCLUSION

The discussion on agricultural economy of Marathwada region help us to draw the following conclusions:

Agriculture is one of the most important sectors of Marathwada region. In the year 2001 about 76.72 per cent of the workers in this region were engaged in agriculture. According to 1992-93 data, about 78 per cent of the workers were engaged in agriculture which reveals that the major portion of the population of this region is dependent on agriculture.

Ecological and geo-chemical atmosphere is favourable to agriculture in Marathwada region. Godavari, Purna and

Manjra are the important rivers of this region. The soils vary greatly in texture and depth. In the years of favourable rainfall bumper rabi crops are grown in these areas. The major portion of the region is also covered by medium black soil. The medium and deep black soil in the region is rich in plant nutrients and can support good kharif and rabi crops like jowar, bajra, cotton, groundnut, wheat, tur, gram, safflower, sunflower and soyabean. The deep black soil covers about 12.95 per cent portion of the Marathwada region. The medium black soil covers about 64.79 per cent and the coarse and shallow soil covers about 22.26 per cent portion of the Marathwada region in the year 2001.

The average annual rainfall is not uniform in all the districts of Marathwada region. Marathwada region receives an average annual rainfall of 827 mms.

According to the population census of 2001 the total population of the region is 155.89 lakhs and about 75.41 per cent of total population lives in rural areas and 24.59 per cent in urban areas.

In Marathwada region out of the total geographical area of 64,43,400 hectares about 3.51 per cent (2,26,100 hectares) has been classified as forest area under land utilization statistics of 1999-2000 maintained by the Agriculture Department. The major portion of the total geographical area i.e. 71.85 per cent is the net area sown. The average size of land holdings in the Marathwada region has declined from 6.18 hectares in 1970-71 to 2.39 hectares in 1990-91. The cropping pattern in almost all the districts is dominated by foodcrops.

Major crops grown in the region are the kharif jowar, bajra, tur, cotton, sugarcane, groundnut, wheat, rabi jowar, gram, safflower, sunflower, sesamum and newly introduced crop soyabean alongwith other crops are also grown in Marathwada region.

In the Marathwada region area under foodgrains was 72.43 per cent in the year 1981-82 which has decreased to 64.12 per cent in the year 1999-2000. The percentage area under cotton in all the districts in Marathwada region has continuously increased during the period 1981-2000. The percentage area under sugarcane and oilseed crops has slightly increased from 1.27 per cent and 11.47 per cent in 1981-82 to 2.08 per cent and 12.78 per cent in the year 1999-2000 respectively. The percentage area under groundnut has continuously declined from 2.23 per cent in 1981-82 to 1.50 per cent in 1999-2000 in Marathwada region. The percentage area under safflower has declined from 4.93 per cent in 1981-82 to 3.43 per cent in 1999-2000. But the percentage area under sunflower and soyabean has increased from 0.69 per cent and 0.007 per cent in 1981-82 to 5.13 per cent and 0.52 per cent in the year 1999-2000 respectively. In short, the percentage area under oilseeds has continuously increased during the period 1981-2000. Lastly, the percentage of total area irrigated to total cropped area in Marathwada region has continuously increased. It has increased from 10.59 per cent in the year 1981-82 to 14.23 per cent in the year 1999-2000.

Marathwada is one of the backward regions of the Maharashtra state. The rate of agricultural mechanization in the region is slow due to low crop intensity along with unawareness of advanced technology among the farmers and also the knowledge towards the use of unconventional energy sources. Marathwada is industrially backward because the use of raw materials locally available are not optimally used for industrial purpose. Low literacy group is also one of the reasons of backwardness of Marathwada region. Even after many years planning neglect of authorities to implement better growth plans and lack of attention towards the socio-economic problems of the region still persist. Hence, more and more encouragement in the direction of opening agro-based and bio-technological units are required in Marathwada region which will promote agricultural, industrial development and economy of the Marathwada region.

REFERENCES

Census of India 2001, Provisional population totals, Controller of Publications, Govt. of India, Delhi.

At A Glance–Aurangabad Division, Office of Economics and Statistics, Aurangabad, 2002.

Brahme Sulabha, Pore Kumud and Pore S. H. (1975), Regional Planning-A Case Study of Marathwada Region, Gokhale Institute of Politics and Economics, Poona.

Brahme Sulabha, Pore Kumud and Pore S.H.: *Ibid.*

Sable, M. S. (1991), Supply Response of some Agricultural Commodities in Marathwada, Unpublished Ph.D. Thesis, Marathwada University, August.

Hedgire, D. N. (1992), A Study of Production, Marketing and Processing of Major Edible Oilseeds in Marathwada-An Economic Analysis. Unpublished. Ph.D. Thesis, Marathwada Agricultural University, Parbhani.

Patil K. K. (1997), Impact of Price and Non-price Factors on Sugarcane Acreage in Marathwada, Unpublished Ph.D. Thesis, Dr. Babasaheb Ambedkar Marathwada University, Aurangabad, October.

Marathwada at a Glance: A Case Study, conducted by Marathwada Agricultural University, Parbhani, 1996.

Epitome of Agriculture in Maharashtra Part II, 2001-02.

4 Growth Performance of Oilseed Crops in Marathwada Region

India holds an important position in the world in terms of rich diversity of oilseed crops and in terms of area under it. India being a major producer of oilseeds has about 23 million hectares of land under oilseed cultivation. India accounts for 10 per cent of oilseed production for which uses 20 per cent of world land under oilseeds. Oilseeds constitute the second largest crop after foodgrains in India. Oilseeds together account for 10.14 per cent of the gross cropped area[1].

Groundnut, sunflower, safflower, sesame, linseed, rapeseed and mustard, castor, nigerseed and soyabean are the important oilseeds at the national level. Andhra Pradesh, Karnataka, Maharashtra, Tamil Nadu, Gujarat, Uttar Pradesh, Madhya Pradesh and Rajasthan are the major oilseed growing states in India[2].

India is among the largest vegetable oil economy in the world accounting for about 14 per cent of the world's oilseeds and 8 per cent of oilseeds production. The country ranks first in the world in the production of castor, safflower, sesame and nigerseed; second in groundnut and rapeseed, mustard; third in linseed and fifth in soyabean and sunflower. However, the productivity of oilseed crops in India is one of the lowest in the world except in case of castor[3]. Production of oilseeds increased from 8.7 million tonnes in 1980-81 to 21.6 million tonnes in the year 2001-2002[4].

The groundnut is an important oilseed crop in India. However, with the growing realization of the nutritive that groundnut kernel possesses, it is now being realized that in times to come groundnut may be regarded as an important food crop also. In the international arena, with an annual production of 6.2 million tonnes in 2001, India contributed 17.7 per cent to the world groundnut production and was ranked the second largest groundnut producing nation after China. India also happens to be one of the major groundnut exporting country besides Argentina, China and the U.S.[5]. India occupies a premier position in the world in safflower accounting for 50 per cent and 40 per cent of area and production, respectively. However, the productivity in India is only about 80 per cent of the world's productivity. Other important countries producing safflower are USA, Mexico, Ethiopia, Australia, China and Argentina. The average productivity in USA and China is more than 2 tonnes per hectare. Despite the premier position India enjoys in the world, safflower plays a minor role in the national vegetable oils scenario accounting for just 1.8 per cent of oilseeds and contributing only 1.1 per cent of oilseeds production in 2000-01[6].

Sunflower introduced just 30 years ago for commercial production in India has emerged as an important edible oilseed crop. The annual planting area under sunflower is around 2 million hectares with a production of 1.2 million tonnes and a productivity of around 600 kg/ha, contributing nearby 4 per cent to the total edible oilseed production in the country. The crop is grown in two main seasons - kharif and rabi. Sunflower is largely grown as rainfed crop particularly in the southern states. The productivity of sunflower in India is of the lowest (584 kg/ha) compared to the world average of 1170 kg/ha and highest productivity of 2200 kg/ha in France in the year 2002-03. Obviously the crop is not performing to its best in India, being restricted to rainfed and of low fertile soils, it is subjected to a variety of stresses resulting in low productivity[7].

Sesame is the most ancient and traditional crop of India. However, the productivity of sesame is low with very wide variations. The reason of low productivity are the cultivation of sesame in marginal and sub-marginal, rainfed conditions with little or no inputs and poor crop management. Sesame crop suffers due to biotic as well as abiotic stress[8].

Linseed is one of the important rabi oilseed crop in India. On global scenario India ranks first in area and third in production. In India, linseed is under cultivation since prehistoric time. It occupies about 8 lakh hectares area and total annual production is about 3 lakh tonnes. Under normal sown conditions and on optimum inputs linseed can yield 16-18 quintals per hectare[9].

Castor known as castrobean is native of Ethiopia. It is traditionally a long duration crop, flowering later and maturing under residual moisture conditions. In India, it is cultivated under rainfed conditions in the states of Andhra Pradesh, Karnataka, Tamil Nadu and Orissa[10].

In India, rapeseed-mustard group of crops play an important role after groundnut and occupy place largely on marginal and sub-marginal lands. However, situation has changed dramatically in recent times. The irrigated area increased steadily from 14.8 per cent in 1967-68 to 69.1 per cent during the years 1996-97 and came down to 58.3 per cent during the year 1998-99. Thus, about 40 per cent of the area under rapeseed mustard is still rainfed. These crops often have to face a number of abiotic and biotic stresses[11].

Soyabean, ranking second among oilseed crops of India, has exhibited unparallel growth in area in the past three decades. The area under the crop is stabilizing now. Presently it covers about 6 million hectares. During these years, the pace of growth in productivity did not commensurate with its horizontal expansion. The national productivity level achieved so far is 1 tonne per hectare. The productivity potential of soyabean varieties developed in recent past is about 4 tonnes per hectare. Under real farm conditions, many farmers are harvesting about 3 tonnes per hectare. If we take productivity levels, about 2 tonnes per hectare of frontline demonstrations as achievable estimate, we are at present harvesting only 50 per cent of the potential productivity of soyabean crop in India. Besides socio-economic status of farmers and lack of awareness on production technology, occurrence of several abiotic and biotic stresses at various stages of crop growth is an important factor responsible for poor productivity levels of soyabean in India. These factors became more important on account of growing crop under rainfed conditions. Moreover, the varieties developed under Indian conditions have a narrow genetic base, which do not permit to evolve newer varieties with higher genetic potential than those existing (2001). Further the tropical growing conditions and the photosensitive nature of crop limit its productivity[12].

India's achievement in the oilseeds sector so far has been quite commendable. It has witnessed a compound growth rate of 2 per cent till 1987 which has shot up by 10 per cent thereafter[13].

India is the largest producer of oilseeds in the world. There has been considerable importance to this commercial crop during the last two decades. In the year 2003-04 at all India level, the area under groundnut was 59.98 lakh hectares and production 81.82 lakh tonnes with an average yield of 1364 kg per hectare. Area under sunflower was 20 lakh hectares and production 9.92 lakh tonnes with an average yield of 496 kg per hectare. Area under safflower was 3.49 lakh hectares and production 1.28 lakh tonnes with an average yield of 367 kgs per hectare. Area under soyabean was 64.97 hectares and production 78.63 lakh tonnes with an average yield of 1210 kg per hectare in the year 2003-04 in India[14].

Maharashtra is one of the leading oilseed growing states in India. The area under groundnut in the Maharashtra state was 3.9 lakh hectares and have total production of 4.53 lakh tonnes with an average yield of 1162 kg per hectare in the year 2003-04. Area under sunflower was 2.93 lakh hectares and production 1.16 lakh tonnes with an average yield of 396 kg per hectare in the year 2003-04. Area under safflower was 2.34 lakh hectares and production was 0.81 lakh tonnes with an average yield of 346 kg per hectare in the year 2003-04. And area under soyabean was 15.89 lakh hectares and production 22.19 lakh tonnes with an average yield of 1396 kg per hectare in the year 2003-04 in Maharashtra[15].

Marathwada is one of the regions of Maharashtra state. Production of groundnut, sunflower, safflower, castorseed, linseed, sesamum, nigerseed, rape and mustard and soyabean is taken in this region. But out of them groundnut, sunflower, safflower, and newly introduced soyabean these four oilseed crops are considered as major oilseeds in this region for the study. Area under total oilseed crops has declined from 7.84 lakh hectares in 1980-81 to 6.62 lakh hectares in the year 1999-2000. But production and yield have increased from 2.69 lakh tonnes and 342 per hectares in the year 1980-81 to 3.74 lakh hectares, 565 kg per hectares in the year 1999-2000 respectively.

Area under groundnut has declined from 1.62 lakh hectares in 1980-81 to 0.86 lakh hectares in the year 1999-2000 in Marathwada region. Area under safflower has declined from 2.90 lakh hectares in 1980-81 to 1.70 lakh hectares in the year 1999-2000. Area under sunflower has increased from 0.42 lakh hectares in 1980-81 to 2.65 lakh hectares in the year 1999-2000 and area under soyabean has increased from 22 hundred hectares in 1990-91 to 32.2 thousand hectares in the year 1999-2000 in Marathwada region.

Production of groundnut has declined from 60,900 tonnes in 1980-81 to 59,700 tonnes in the year 1999-2000 in Marathwada region. Production of safflower has declined from 1,42,400 tonnes in 1980-81 to 92,400 tonnes in the year 1999-2000. Production of sunflower has increased from 35,800 tonnes to 1,39,800 tonnes in the year 1999-2000. Production of soyabean has increased from 1700 tonnes in 1990-91 to 48,800 tonnes in the year 1999-2000 in Marathwada region.

Average yield for groundnut and safflower has increased from 375 kg per hectare and 492 kg per hectare in the year 1980-81 to 696 kg per hectare and 544 kg per hectare in the year 1999-2000 respectively in Marathwada region. Yield of sunflower has declined from 852 kg per hectare in 1980-81 to 528 kg per hectare in the year 1999-2000 and average yield of soyabean has increased from 773 kg per hectare in 1990-91 to 1510 kg per hectare in the year 1999-2000 in Marathwada region.

The present chapter aims at to study growth performance of oilseed crops i.e. groundnut, safflower, sunflower and soyabean in the different districts of the Marathwada region and the region as a whole. Annual compound growth rates of area, production and yield of groundnut, safflower, sunflower and soyabean for different districts of Marathwada are computed. The study period is divided into three sub-periods. Sub-period I from 1980-81 to 1989-90, sub-period II comprises ten years from 1990-91 to 1999-2000 and overall period comprises twenty years from 1980-81 to 1999-2000.

For calculating annual compound growth rates the exponential function of the form $y = ab^x$ was used which becomes a semi-log function of the form log y = log a + x log b for the operational purpose. Here y stands for area, production and yield. 'a' and 'b' are the constants and 'x' is the time factor. (b-1)´ 100 gives compound growth rate in percentage.

RESULTS AND DISCUSSION

Groundnut

Table 1 presents districtwise area, production and yield growth rates of groundnut in Marathwada region. It is clear from the table that the area growth rate for sub-period I is positive (1.8 per cent) but for sub-period II and for overall period it is negative. It means that the area under groundnut in Marathwada region as a whole is declining. District level pictures of growth rate of groundnut reveals that the area growth rates were positive in Beed (3.27), Latur (3.91) and very high in Parbhani district (8.83) and in case of remaining districts it was negative in sub-period I.

If we observe the growth rates of area in sub-period II we conclude that growth rates of groundnut are negative in all the districts of Marathwada and Marathwada region as a whole. It is clear that all the districts had lost area under groundnut cultivation during this sub-period. For overall period (1980-2000), the growth rates of groundnut area were negative in most of the districts and in Marathwada region as a whole (-1.68 per cent) but the growth rate of this crop is positive (1.3 per cent) in Beed district and it is also comparatively less than sub-period I. The growth rates of area of groundnut show that the area under groundnut

Table 1
Districtwise Annual Compound Growth Rates of Area, Production and Yield of Groundnut in Marathwada Region

(in per cent)

	Sub-period I (1980-81 to 1989-90)			Sub-period II (1990-91 to 1999-2000)			Overall period (1980-81 to 1999-2000)		
Districts	*Area*	*Production*	*Yield*	*Area*	*Production*	*Yield*	*Area*	*Production*	*Yield*
Aurangabad	-8.06	-9.51	-1.58	-2.67	-2.11	0.58	-2.23	0.86	3.15
Jalna	-0.78	1.56	0.80	-11.68	-13.55	-2.12	-5.4	-4.27	1.37
Beed	3.27	5.85	2.50	-7.38	-8.85	-1.69	1.3	1.37	0.09
Osmanabad	-4.83	-1.38	3.62	-4.55	-7.58	-3.16	-3.04	-4.45	-1.45
Latur	3.91	-0.73	-3.97	-6.02	-5.45	0.61	-2.86	-2.25	1.19
Parbhani	8.83	15.37	6.01	-14.59	-16.73	-2.51	-0.74	1.69	2.45
Nanded	-0.09	6.65	6.74	-8.86	-11.52	-2.92	-3.1	-0.33	2.86
Marathwada	1.8	6.42	0.63	-8.08	-10.63	0.01	-1.68	-0.17	0.24

Note: Compound growth rates for sub-period I of Jalna and Latur districts are calculated from 1981-82.
Source: Compound growth rates are based on the data collected from (i) Season and crop reports (ii) Epitome of Agriculture in Maharashtra, Part-II and (iii) Socio-Economic Review and District Statistical Abstracts.

crop is declining day by day in all the districts of Marathwada region.

The growth rate of area under groundnut for Marathwada region has decreased from 1.8 per cent in sub-period I to -8.08 per cent in sub-period II. District level picture reveals that the growth rates of area for Aurangabad and Osmanabad districts remained negative but the growth rates of area for Jalna, Beed, Latur, Parbhani and Nanded districts have decreased from -0.78 per cent, 3.27 per cent, 3.91 per cent, 8.83 per cent and -0.09 per cent respectively in sub-period I to -11.68 per cent, -7.38 per cent, -6.02 per cent, 14.59 per cent and -8.86 per cent respectively in sub-period II.

Table 1 indicates that the production of groundnut has shown a negative growth (-0.17 per cent per annum) during the overall period in Marathwada region even though there was a positive growth in the yield of this crop. This was due to negative growth in the area under this crop. The districtwise growth performance of this crop during the overall period has also been disappointing in most of the districts of Marathwada region. Production of groundnut has recorded negative growth in Jalna, Osmanabad, Latur and Nanded districts. However, production of groundnut is found to have grown at a positive rate of 1.69 per cent, 1.37 per cent and 0.86 per cent in Parbhani, Beed and Aurangabad districts respectively. Groundnut production performance in Parbhani district is sound as compared to other districts in Marathwada region during the overall period.

In Marathwada region, growth rate of groundnut production has decreased from 6.42 per cent in sub-period I to -10.63 per cent in sub-period II. Growth performance of this crop is positive in Nanded, Parbhani, Beed and Jalna districts during sub-period I but for sub-period II growth performance of groundnut production is negative in all the districts.

The growth rate of yield is found to be positive in Aurangabad, Jalna, Beed, Latur, Parbhani and Nanded districts and in Marathwada region as a whole during the overall period but in case of Osmanabad district it is negative (-1.45 per cent). Growth rate of yield was positive but decreased from 0.63 per cent in sub-period I to 0.01 per cent in sub-period II in

Marathwada region as a whole. Growth rate of yield increased in Aurangabad district from –1.58 per cent in sub-period I to 0.58 per cent in sub-period II but in rest of the districts it decreased during the same sub-period.

Safflower

Table 2 indicates the period-wise growth rates of area, production and yield of safflower in Marathwada region. It can be seen from the table that all the districts except Aurangabad registered positive growth rate of area along with Marathwada region as a whole during the sub-period I. The highest growth rate of area in this sub-period is in Nanded district (20.57 per cent) and lowest but negative in Aurangabad district (-1.98 per cent). It means that the area under safflower crop increased in Marathwada region in sub-period I. Production rise was found in Nanded, Parbhani, Osmanabad and Beed districts which came as the result of positive yield improvement in this period and decrease in production was noted in Jalna and Latur districts which was due to the reduction in yield while in Aurangabad district production declined due to fall in area in this period. In Marathwada region production of safflower increased due to increase in area under safflower despite increase in yield in this period.

In sub-period II, area, production and yield rates have recorded negative growth but there are some improvements in yield in Marathwada region as a whole. For this period growth rate of area under safflower is the highest in Latur district (4.12 per cent) and the lowest and negative in Aurangabad district (-8.24 per cent). The growth rate of area under safflower for Jalna, Beed, Osmanabad and Parbhani districts are –5.68 per cent, -2.26 per cent, 0.66 per cent, -4.22 per cent and 0.53 per cent respectively in this sub-period. The production performance of this crop was better in Latur district only. Production growth was negative in Nanded, Parbhani, Osmanabad, Beed, Jalna and Aurangabad districts in the sub-period II. Yield growth rates of safflower were negative in all the districts. Production of safflower is found to have grown at a positive rate of 4.06 per cent due to positive increment in

Table 2
Districtwise Annual Compound Growth Rates of Area, Production and Yield of Safflower in Marathwada Region

(in per cent)

	Sub-period I (1980-81 to 1989-90)			Sub-period II (1990-91 to 1999-2000)			Overall period (1980-81 to 1999-2000)		
Districts	*Area*	*Production*	*Yield*	*Area*	*Production*	*Yield*	*Area*	*Production*	*Yieid*
Aurangabad	-1.98	-2.10	-0.12	-8.24	-10.05	-2.01	-5.49	-5.79	-0.32
Jalna	13.96	-8.22	-6.11	-5.68	-8.29	-2.73	-1.24	-6.64	-2.31
Beed	1.13	7.61	6.71	-2.26	-3.81	-1.59	-2.94	-4.66	-1.42
Osmanabad	3.68	8.00	4.23	0.66	-2.22	-2.84	0.75	-1.12	-1.83
Latur	2.61	-1.59	-3.98	4.12	4.06	-0.05	1.16	0.76	-0.40
Parbhani	2.59	11.84	9.00	-4.22	-8.62	-4.59	0.18	-1.62	-1.80
Nanded	20.57	27.21	6.54	0.53	-0.80	-1.25	9.12	8.62	0.22
Marathwada	3.45	5.07	-0.18	-3.29	-6.20	-0.06	-1.20	-3.07	0.03

Note: Compound growth rates for sub-period I of Jalna and Latur districts are calculated from 1981-82.

Source: Compound growth rates are based on the data collected from (i) Season and crop reports (ii) Epitome of Agriculture in Maharashtra Part-II and (iii) Socio-Economic Review and District Statistical Abstracts.

area of safflower in Latur district. In this sub-period there has been negative growth in the yield of this crop in all the districts of Marathwada.

The growth rate of production of safflower for Marathwada region has decreased from 5.07 per cent in sub-period I to -6.20 per cent in sub-period II. Production rate of safflower in Latur district has increased from –1.59 per cent in sub-period I to 4.06 per cent in sub-period II. But growth rates of production of safflower in Aurangabad, Jalna, Beed, Osmanabad, Parbhani and Nanded districts have decreased from –2.10 per cent, -8.22 per cent, 7.61 per cent, 8.00 per cent, 11.84 per cent and 27.21 per cent respectively in sub-period I to –10.05 per cent, -8.29 per cent, -3.81 per cent, -2.22 per cent, -8.62 per cent and –0.80 per cent in sub-period II.

Table 2 reveals that the production of safflower has shown a negative growth (-3.07 per cent) during the overall period (1980-81 to 1999-2000) in Marathwada region even though there was a positive growth in the yield of safflower crop. This was due to negative growth in the area under safflower crop. The districtwise picture reveals that the growth performance of safflower crop was disappointing in Aurangabad, Jalna, Beed, Osmanabad, Parbhani and Latur districts in Marathwada region during the overall period, because production of safflower has recorded negative growth in these districts. However, production of safflower is found to have grown at a positive rate of 0.76 per cent in Latur district only.

For the overall period, the growth rate of area of safflower in the Marathwada region is found to be negative (i.e. –1.20 per cent). For this period, the growth rate of area of safflower is the highest in Nanded district (i.e. 9.12 per cent) and the lowest but negative in Aurangabad district (i.e. –5.49 per cent) and for Jalna, Beed, Osmanabad, Latur and Parbhani districts the growth rates are –1.24 per cent, -2.94 per cent, 0.75 per cent, 1.16 per cent and 0.18 per cent, respectively.

Sunflower

The districtwise annual compound growth rates of area, production and yield of sunflower crop in Marathwada region are presented in Table 3. It can be noticed from the table that Marathwada region has positive rise in area, production and yield during the sub-period I. This production growth was the result of area expansion in the region. The production of sunflower has shown a positive growth (28.68 per cent per annum) during the sub-period I (1980-1990) due to the positive growth in the yield of sunflower in Marathwada region. The districtwise growth performance of sunflower crop during the sub-period first has also been better in all the districts of the Marathwada region. The production of sunflower is found to have grown at a positive rate of 68.58 per cent, 56.22 per cent, 54.78 per cent, 49.30 per cent, 45.76 per cent, 13.10 per cent and 13.06 per cent in Parbhani, Jalna, Nanded, Aurangabad, Beed, Latur and Osmanabad districts respectively. Out of this, production growth is higher in Parbhani district and lower in Osmanabad district. But the growth rates of yield of sunflower in all the districts are found to be negative. The growth rate of area of sunflower is the highest in Parbhani district (76.47 per cent) and the lowest in Osmanabad district (22.37 per cent). It means that the area, production and yield growth is less in Osmanabad district as compared to other districts during the first sub-period.

From the Table 3, it seems that positive area increase was noticed in the sub-period II in the Marathwada region. Growth of area of sunflower crop is highest in the Jalna district (6.15 per cent) and it registered lowest growth in Nanded district (-1.35 per cent) in this sub-period. Loss of area was witnessed in Beed, Parbhani and Nanded districts. In other districts there was positive growth of area but less as compared to sub-period I. In this period sunflower area growth is positive but very negligible in Marathwada region.

As far as production growth is concerned during the sub-period II, Jalna and Aurangabad districts have recorded positive production growth that came mainly through positive area expansion during this sub-period II. Production growth performance in Beed, Osmanabad, Latur, Parbhani and Nanded districts and Marathwada region as a whole showed negative production growth in this period. Yield growth rates

Table 3
Districtwise Annual Compound Growth Rates of Area, Production and Yield of Sunflower in Marathwada Region

(in per cent)

	Sub-period I (1980-81 to 1989-90)			Sub-period II (1990-91 to 1999-2000)			Overall period (1980-81 to 1999-2000)		
Districts	*Area*	*Production*	*Yield*	*Area*	*Production*	*Yield*	*Area*	*Production*	*Yield*
Aurangabad	49.95	49.30	-1.78	4.18	1.56	-2.50	26.12	27.05	0.15
Jalna	73.76	56.22	-4.04	6.15	3.14	-2.84	24.89	22.16	-0.88
Beed	48.45	45.76	-1.65	-1.27	-2.25	-0.98	15.06	17.69	2.18
Osmanabad	22.37	13.06	-7.60	0.55	-1.88	-2.39	9.46	10.14	0.63
Latur	26.57	13.10	-7.42	0.46	-1.22	-1.57	7.66	9.00	2.02
Parbhani	76.47	68.58	-3.14	-0.52	-0.62	0.12	29.36	30.72	1.79
Nanded	62.66	54.78	-4.56	-1.35	-5.56	-3.80	24.18	26.92	2.31
Marathwada	34.37	28.68	0.81	0.29	-1.45	-0.10	12.82	14.40	-0.06

Note: Compound growth rates for sub-period I of Jalna and Latur districts are calculated from 1981-82.

Source: Compound growth rates are based on the data collected from (i) Season and crop reports (ii) Epitome of Agriculture in Maharashtra Part-II and (iii) Socio-Economic Review and District Statistical Abstracts.

also suggest that there were fluctuations in yield growth of sunflower in Marathwada region in the sub-period II. There was stagnancy in almost all the districts except in Parbhani district that registered a positive growth in yield of sunflower during the sub-period II. The districtwise growth performance of this crop during the sub-period II has also been disappointing in all the districts of Marathwada region.

It is clear from the Table 3 that the production of sunflower has shown a positive growth (14.40 per cent) during the overall period even though there was a negative growth in the yield of this crop. This was due to positive growth in the area under this crop in Marathwada region as a whole. The growth rates of area of sunflower for Aurangabad, Jalna, Beed, Osmanabad, Latur, Parbhani and Nanded districts are 26.12 per cent, 24.89 per cent, 15.06 per cent, 9.46 per cent, 7.66 per cent, 29.36 per cent, and 24.18 per cent respectively. It means that the growth performance of area is positive in all the districts of the Marathwada region during the overall period. But better area growth performance of this crop is found in Parbhani district.

Production of sunflower is found to have grown at a positive rate of 30.72 per cent, 27.05 per cent, 26.92 per cent, 22.16 per cent, 17.69 per cent, 10.14 per cent, and 9 per cent in Parbhani, Aurangabad, Nanded, Jalna, Beed, Osmanabad and Latur districts respectively. From this it is obvious that the districtwise growth performance of this crop is better during the overall period.

Table 3 indicates that the production of sunflower has shown a positive growth during the overall period even though there was a negative growth in the yield of this crop in Jalna district.

Soyabean

Area, production and yield growth rates of soyabean in Marathwada region are shown in Table 4. Growth rates were worked out for those periods for which regular data was available. Growth rates for overall period refers to 1993-94 to 1999-2000. It can be seen from the table that the area and production growth is positive in Marathwada region but yield growth is zero. Production growth is found to be positive due to the positive growth in area. The district level picture reveals that the growth rates of area of soyabean crop are positive for Nanded, Parbhani, Osmanabad and Jalna districts and negative for Aurangabad, Beed and Latur districts. But the districtwise growth performance of this crop during the overall period has also been disappointing in most of the districts of Marathwada. Production of soyabean has recorded negative growth in Aurangabad and Beed districts. However, production of soyabean is found to have grown at a positive rate of 27.63 per cent, 26.88 per cent, 21.87 per cent, 4.64 per cent and 2.08 per cent in Nanded, Parbhani, Osmanabad, Jalna and Latur districts respectively. There was a negative growth in the yield of this crop in Jalna and Osmanabad districts but in rest of the districts it is found to be positive.

Table 4
Annual Compound Growth Rates of Area, Production and Yield of Soyabean in Marathwada Region

(in per cent)

	Period (1993-94 to 1999-2000)		
Districts	*Area*	*Production*	*Yield*
Aurangabad	-7.27	-9.91	3.71
Jalna	9.37	4.64	-1.85
Beed	-13.42	-16.66	1.08
Osmanabad	32.87	21.87	-5.73
Latur	-1.13	2.08	3.52
Parbhani	22.50	27.63	6.63
Nanded	23.97	26.88	5.18
Marathwada	6.11	12.00	0.00

Source: Compound growth rates are based on the data collected from (i) Season and crop reports, (ii) Epitome of Agriculture in Maharashtra Part-II and (iii) Socio-Economic Review and District Statistical Abstracts.

CONCLUSION

Oilseed crop is the most ancient crop of India. India is the largest producer of oilseeds in the world economy. Oilseeds play dominant role in the economic development of the country and it is an important base for the agro-industries in India. During the last two decades there has been considerable importance to this crop due to the higher growth rate of population.

India is among the largest vegetable oil economies in the world accounting for 14 per cent of the world's oilseeds area and 8 per cent of oilseeds production. The country ranks first in the world in the production of castor, safflower, sesamum and niger seed; second in groundnut, rape and mustard; third in linseed and fifth in sunflower and soyabean. However, the productivity of oilseed crops in India is one of the lowest in the world except in case of castor.

Now, let us list some important findings of the district level analysis of growth in area, production and yield of groundnut, safflower, sunflower and soyabean crops in different districts of Marathwada and Marathwada region as a whole. Groundnut, which is a pre-eminent crop of the region, production has shown a negative growth during the overall period, even though there was a positive growth in the yield of this crop. This was due to negative growth in the area under this crop. District level analysis of the growth rates revealed that there has been a loss of area under groundnut cultivation in most of the districts but even then production growth performance of it was better in Aurangabad and Parbhani districts during the overall period. Yield growth performance of this crop was comparatively better in Aurangabad, Parbhani and Nanded districts. Area, production and yield growth rates are positive in Beed district and negative in Osmanabad district during the overall period

In case of safflower also the production has shown a negative growth even though there was a positive growth in the yield of this crop. This was due to negative growth in the area under this crop during the overall period in Marathwada region as a whole. Safflower production performance was better in Nanded and Latur districts and in rest of the districts it is negative. Yield growth performance of this crop was better in Nanded district only. It was observed that production of this crop was influenced by area changes in the districts cultivating this crop.

Sunflower production growth performance was better in all the districts of Marathwada region during the overall period, even though there was a negative growth in the yield of this crop. The district level picture revealed that the area production and yield growth rates are positive in all the districts except Jalna district where the growth rate of yield is found to be negative.

Production growth performance of soyabean was better due to the expansion of area under soyabean in Marathwada region as a whole. But the district level picture showed that the growth performance of production of this crop was better in most of the districts except Beed and Aurangabad districts. Yield growth rates are also positive in most of the districts except Osmanabad and Jalna districts of the Marathwada region.

NOTES AND REFERENCES

1. *CMIE*, Monthly Review of the Indian Economy, Economic Intelligence Service, Center for Monitoring Indian Economy Pvt. Ltd., Mumbai, November, 2001.
2. *Ibid.*
3. Hegde, D. M. (2002), Measures to Turn Self-reliant, The Hindu Survey of Indian Agriculture.
4. Chand Ramesh, Jha Dayanath and Mittal Surbhi (2004), WTO and Oilseeds Sector: Challenges to Trade, Liberalization, *Economic and Political Weekly*, February 7.
5. Basa, M. S. (2003), Stress Management in Groundnut, National Seminars on Stress Management in Oilseeds for Attaining Self Reliance in Vegetable Oils, Indian Council of Agricultural Research, Rajendranagar, Hyderabad, January 28-30.
6. Hegde, D. M. (2003), Stress Management in Safflower, National Seminar on Stress Management in Oilseeds for Attaining Self Reliance in Vegetable Oils, Indian Council of Agricultural Research, Rajendranagar, Hyderabad, January 28-30.
7. Seetharam, A. (2003), Stress Management in Sunflower, National Seminars on Stress Management in Oilseeds for Attaining Self Reliance in Vegetable Oils, Indian Council of Agricultural Research, Rajendranagar, Hyderabad, January 28-30.

8. Duhoon, S. S. (2003), Stress Management in Sesame and Niger, National Seminar on Stress Management in Oilseeds for Attaining Self Reliance in Vegetable Oils, Indian Council of Agricultural Research, Rajendranagar, Hyderabad, January 28-30.

9. Kerkhi, S. A. (2003), Stress Management in Linseed, National Seminar on Stress Management in Oilseeds for Attaining Self Reliance in Vegetable Oils, Indian Council of Agricultural Research, Rajendranagar, Hydria, January 28-30.

10. Hanumantha Rao, C. and Lavanya, C. (2003), Stress Management in castrobean, National Seminar on Stress Management in Oilseeds for Attaining Self Reliance in Vegetable Oils, Indian Council of Agriculture Research, Rajendranagar, Hyderabad, January 28-30.

11. Arvind Kumar and Premi, O. P. (2003), Stress Management in Rapeseed-Mustard, National Seminar on Stress Management in Oilseeds for Attaining Self Reliance in Vegetable Oils, Indian Council of Agriculture Research, Rajendranagar, Hyderabad, January 28-30.

12. Joshi, O. P. and Bhatia, V. S. (2003), Stress Management in Soyabean, National Seminar on Stress Management in Oilseeds for Attaining Self Reliance in Vegetable Oils, Indian Council of Agriculture Research, Rajendranagar, Hyderabad, January 28-30.

13. Singh Parmatma and Srivastava Alka (1995), Oilseed Production in India: Changing Proportions, *Agricultural Situation in India*, Vol. XLIX, No. 12.

14. *CMIE*, Monthly Review of the Indian Economy. Economic Intelligence Service, Centre for Monitoring Indian Economy Pvt. Ltd., Mumbai, March, 2005.

15. *CMIE, Ibid.*

5 Acreage Response of Oilseed Crops in Marathawada Region

In a developing country like India the policy directed towards speeding up the growth of agricultural sector becomes crucial one. In the dynamic context, production of subsistence crops seems to be insufficient to meet the growing needs of the rapidly rising population which necessitates the growth and expansion of agricultural sector. The growth of agricultural sector depends upon the agricultural production. The growth of agricultural production depends on various factors such as cropping pattern, weather, rainfall, irrigation, farm prices and technology, etc. Apart from non-price factors, price is one of the most important factors affecting agricultural production in particular and agricultural growth in general. Recently, it has been generally observed that output of a crop responds in the short run to price incentives and in the long run to technological improvements. At a given level of agricultural technology, an increase in the price of a particular commodity stimulates a reallocation of area in favour of that crop and thereby raises its production. Adoption of appropriate technological and organizational changes are important to achieve agricultural growth. Whenever a new technology penetrates traditional agriculture, the production function shifts and even at the same time price level of agricultural output registers substantial increase. Thus, agricultural growth requires sound agricultural policy.

Agricultural price policy is one of the important instruments in the hands of the government to achieve higher growth rates in agricultural production in order to make price policy an effective instrument of inducing desired changes in resource allocation to different crop enterprises, knowledge of farmers supply responsiveness is essential. The farmer is a central point who makes the final decision concerning the allocation of land to different crops depending on his expected revenue from different crops. The crucial decision relating to what crop and how much crop to grow on how much land with what levels of non-land factors of production are essentially based on the relative profitability of each crop. Any change in the production levels of different crops is thus brought out through inter-crop shifts on land owing to varying profitability from crop to crop. Thus, acreage response of a particular crop is influenced by the possibilities of engaging it in the other lines of production.

Production of agricultural commodities is a function of price and non-price variables. In other words, price alone does not determine farmer's acreage allocation decisions as was assumed under the classical static situation qualified by ceteris paribus assumptions. But non-price factors are also important in determining farmer's acreage allocation decisions. Therefore, it is hypothesized here that acreage allocation decisions are influenced by price and non-price variables. Since the magnitude of farmers response to different factors varies not only from crop to crop but even for the same crop from district to district due to agro-climatic factors, adoption of technology, institutional and other socio-economic factors. Therefore, such factors must be taken into consideration while examining response behaviour.

The question of farmers responsiveness to price signals has been extensively examined by number of researchers. In most of these studies Nerlovian model has been used either as such or with minor modifications. The researchers generally employed two types of models in estimating acreage response of farmers. The first is called as 'Traditional price lag model' and the second is 'Adjustment lag model' also known as 'Nerlovian model'. However, the fact cannot be denied that the price changes do influence farmers decision making regarding the allocation of land resources. Thus, the farmers responsiveness to the price changes is helpful in formulating an appropriate price policy which is of immense importance for the growth and stability of the production on one hand and economy on the other. This is true for formulating policies at the state as well as regional levels. It carries high weightage with respect to oilseed crops in particular and cash crops in general.

Economists have extensively investigated the growth performance of rice and wheat during the last four decades. It has been widely researched at the global, national, state, regional and household levels. Unfortunately, scant attention has been paid to the study of oilseed crops, which play an important role in sustaining crop systems and meeting the domestic needs of edible oil in the country. Therefore, an attempt has been made to provide current evidence on the temporal and spatial dimensions of the oilseed development. The production of oilseeds like other agricultural commodities is primarily determined by acreage. Therefore, it would be worthwhile to investigate the factors influencing acreage. Various price and non-price factors influence the farmer's decisions regarding land allocation to different crops.

In this chapter, an attempt is made to gauge the impact of price and non-price factors on the farmers decision to allocate the land to oilseed crops grown in the Marathwada region. More specially, our object here is to estimate the acreage response of oilseed crops, viz., groundnut, safflower, sunflower and soyabean grown in Marathwada region and to examine the role played by price and non-price factors, viz., price, yield, relative price, relative yield, lagged acreage, irrigation, rainfall, variables explaining the variations in the acreage under oilseed crops studied in different districts of the Marathwada region.

THE DATA AND METHODOLOGY

The present study pertains to oilseed crops, viz., groundnut, safflower, sunflower and soyabean grown in the Marathwada region of Maharashtra state and is based on secondary data at the district level. The required time series data for the period 1980-81 to 1999-2000 on area, production, yield and farm harvest prices of important oilseed crops, gross irrigated area and rainfall during the sowing periods have been collected from various secondary sources published and maintained by the Government of Maharashtra and India such as Season and Crop Reports of Maharashtra state, Districtwise General Statistical Information of Agricultural Department (Epitome of Agriculture, Part I and II), Socio-Economic Review and District Statistical Abstract of different districts of Marathwada region and Farm Harvest Prices in India. For some years, the above mentioned data have been collected from the records maintained by the Directorate of Agriculture, Pune. The present study covers the period from 1980-81 to 2000. This entire period is sub-divided into two periods. The first period relates to eighties beginning from 1980-81 to 1989-90 and the second period relates to nineties from 1990-91 to 1999-2000. These represent the pre- and post-reforms period.

We have computed the yield per hectare of oilseed crops, viz., groundnut, safflower, sunflower and soyabean crop by dividing production by area. We have also computed the average rainfall at different rain-gauge stations in each district for Marathwada as a whole, the average rainfall of seven districts of Marathwada region has been taken. For few years, the price data are not available for the district under study, in that case we have used average price of Marathwada region and minimum support price also. We have also computed the relative price and relative yield of the crops under study. We have computed the weighted price of the competing crops by using proportion of the values of crops in 't' th period as weights. Similarly, the weighted yields of competing crops are computed by using proportion of the values of crops in 't' th period as weights. We have also computed the relative price and yield of groundnut, safflower, sunflower and soyabean crops. RP_t is the relative price of the crop, i.e., the price of the crop deflated by the weighted price of the competing crop in 't' th period. Similarly, 'RY_t' is the relative yield of the crop and deflated by the weighted yield of the competing crop in 't' th period.

While selecting the competing crops, we have considered the relative importance of the crops, relationship between the crops, the type of the crop and period or season in which it is grown. As the competing crop really competes for the same land in the same period we have considered bajra as the competing crop for groundnut in Aurangabad, Jalna, Beed and Osmanabad districts. Kharif jowar is considered as competing crop for groundnut in Latur, Parbhani and Nanded districts and bajra is considered as competing crop for groundnut for Marathwada region as a whole. Rabi jowar is considered as competing crop for safflower in Jalna, Beed and Osmanabad districts. Gram is considered as competing crop for safflower in Aurangabad, Latur, Parbhani and Nanded districts and for Marathwada as a whole. Black gram is considered as competing crop for sunflower in Latur and Nanded districts, Kharif jowar is considered as competing crop for sunflower in Aurangabad district, tur is considered as competing crop for sunflower in Jalna, Beed, Osmanabad and Parbhani districts and for Marathwada region as a whole. Tur is considered as competing crop for soyabean in Aurangabad, Latur and Nanded districts, black gram is considered as competing crop for soyabean in Jalna, Beed, Osmanabad and Parbhani districts and for Marathwada region as a whole.

In the present study, Nerlovian adjustment lag model as a basic framework, the reduced form of the equation is used to assess the farmers acreage response behaviour in respect of oilseed crops, viz., groundnut, safflower, sunflower and soyabean crops grown in the Marathwada region. The acreage response function for oilseeds is specified as follows:

$$A_t = a + b_1A_{t-1} + b_2RP_{t-1} + b_3RY_{t-1} + b_4GI_{t-1} + b_5RF_t + U_t$$

Where

A_t = Area under the crop in the year t (in ha.),

A_{t-1} = Area under the crop in the year t-1 (in ha.),

RP_{t-1} = Relative price of the crop in the year t-1,

RY_{t-1} = Relative yield of the crop in the year t-1,

GI_{t-1} = Gross irrigated area in the year t-1,

RF_t = Sowing period rainfall in the year t

U_t = Error term

Employing the Nerlovian adjustment lag model as the basic frame of analysis by considering sown area as proxy to supply which is the dependent variable and incorporating the lagged acreage, lagged relative price, lagged relative yield, lagged gross irrigated area and sowing period rainfall as the independent variables in different districts of Marathwada region. One year lag is used in acreage, yield, price, irrigation assuming that the current acreage, yield, price and irrigation generally influence the decision about area allocation in the next year. We estimated the acreage response relationship in the study with the help of multiple regression equation. To get the elasticities directly, we estimated the equations using double log or log-log model with all variables in their log form. Logarithmic functions are preferred to linear form because they provide readily estimates of elasticities and also because they yield relatively better results than the linear form. The price elasticity of area planted under a crop is a good approximation to the elasticity of output if it is reasonable to assume that inputs other than land are varied at least in proportion to acreage and returns to scale are not diminishing.

We attempted to estimate acreage response of four oilseed crops, viz., groundnut, safflower, sunflower and soyabean in Marathwada region. The estimated acreage response functions for seven districts and for Marathwada region as a whole along with the value of co-efficient of multiple determination (R^2), the co-efficient of adjustment (B) and standard errors are given in the Tables 1 to 4.

RESULTS AND DISCUSSION

Groundnut

The estimated acreage response coefficients with respect to groundnut for seven districts of Marathwada and for Marathwada region as a whole along with the values of R^2, the co-efficient of adjustment (B) and standard errors are given in the Table 1.

The value of R^2 i.e., the co-efficient of multiple determination is to be 0.754, meaning that all the independent variables together explain 75 per cent of variation in the dependent variable and the remaining 25 per cent variation by the other variables not included in the equation. The coefficient of multiple determination was higher for sub-period II (0.917) than that for sub-period I (0.893).

The regression co-efficient of relative price variable is found to be positive for Jalna, Beed, Parbhani and Nanded districts, but these coefficients are statistically insignificant during the sub-period I. It shows that the relative price of groundnut lagged one year has shown positive but insignificant impact on groundnut acreage allocation decisions of the farmers in the above districts. Similar results were found in the studies of Reddy[1], Jhala[2], Kaul[3], Madhavan[4] and Sable[5]. The coefficients of relative price variables are found to be negative and insignificant in case of Aurangabad, Osmanabad, Latur districts and for Marathwada region as a whole. Jhala[6] had found negative co-efficient of relative price variable in case of regions forming North Gujarat, Marathwada, Madhya Maharashtra, Vidarbha, Karnataka and Tamil Nadu also. For the sub-period II, the coefficient of relative price variable is positive for Aurangabad, Beed and Osmanabad districts and negative for Jalna, Latur, Parbhani, Nanded districts and Marathwada region as a whole. For overall period, the co-efficient of relative price of groundnut is found to be positive but insignificant for Aurangabad, Jalna, Beed, Osmanabad districts and for Marathwada region as a whole and negative and insignificant for Latur, Parbhani and Nanded districts.

The regression coefficients of relative yield variable are found to be positive but insignificant in case of Aurangabad, Jalna, Beed, Osmanabad, Parbhani and Nanded districts and for Marathwada region as a whole. The same result was also found in the studies conducted by Sable[7] and Reddy[8]. They found positive but insignificant co-efficient of relative yield variable in case of groundnut. The regression co-efficient with respect to relative yield is found to be negative in case of Latur district only for the sub-period I. Coefficient of relative yield

Table 1
Estimated Acreage Response Coefficients for Groundnut in Marathwada Region

	Aurangabad			*Jalna*			*Beed*			*Osmanabad*		
	Sub-Period I	*Sub-period II*	*Overall period*	*Sub-period I*	*Sub-period II*	*Overall period*	*Sub-period I*	*Sub-period II*	*Overall period*	*Sub-period I*	*Sub-period II*	*Overall period*
	1980-1981 to 1989-1990	*1990-1991 to 1999-2000*	*1980-1981 to 1999-2000*	*1980-1981 to 1989-1990*	*1990-1991 to 1999-2000*	*1980-1981 to 1999-2000*	*1980-1981 to 1989-1990*	*1990-1991 to 1999-2000*	*1980-1981 to 1999-2000*	*1980-1981 to 1989-1990*	*1990-1991 to 1999-2000*	*1980-1981 to 1999-2000*
Constant in log terms (a)	8.298 (13.380)	9.249 (4.188)	1.594 (1.359)	4.729 (5.663)	-2.261 (2.874)	1.390 (2.172)	0.683 (6.162)	11.520 (5.648)	0.312 (1.437)	0.831 (1.999)	10.835 (5.936)	1.169 (1.754)
Relative price lagged one year (RP_{t-1})	-1.942 (4.328)	0.340 (0.319)	0.165 (0.304)	1.116 (0.905)	-0.628 (0.535)	0.083 (0.403)	0.648 (1.810)	0.078 (0.445)	0.069 (0.445)	-1.388 (1.110)	0.690 (0.519)	0.004 (0.493)
Relative yield lagged one year (RY_{t-1})	0.223 (0.296)	0.008 (0.169)	0.104 (0.121)	0.183 (0.518)	-0.031 (0.222)	-0.176 (0.163)	0.765 (0.886)	0.407 (0.273)	0.162 (0.153)	0.106 (0.102)	0.411* (0.243)	0.134 (0.125)
Gross irrigated area lagged one year (GI_{t-1})	0.146 (0.382)	-0.885 (0.459)	-0.003 (0.117)	0.019 (1.010)	0.331 (0.444)	-0.387 (0.276)	0.162 (0.523)	-1.225 (0.623)	0.085 (0.219)	1.539** (0.432)	-2.376 (1.309)	0.155 (0.359)

Contd...

	Aurangabad			*Jalna*			*Beed*			*Osmanabad*		
	Sub-Period I	*Sub-period II*	*Overall period*	*Sub-period I*	*Sub-period II*	*Overall period*	*Sub-period I*	*Sub-period II*	*Overall period*	*Sub-period I*	*Sub-period II*	*Overall period*
	1980-1981 to 1989-1990	*1990-1991 to 1999-2000*	*1980-1981 to 1999-2000*	*1980-1981 to 1989-1990*	*1990-1991 to 1999-2000*	*1980-1981 to 1999-2000*	*1980-1981 to 1989-1990*	*1990-1991 to 1999-2000*	*1980-1981 to 1999-2000*	*1980-1981 to 1989-1990*	*1990-1991 to 1999-2000*	*1980-1981 to 1999-2000*
Sowing period rainfall (RF_t)	-0.497	-0.031	0.022	0.164	0.081	0.139	-0.145	0.206	0.024	-0.213	0.780*	0.159
	(0.934)	(0.122)	(0.112)	(0.216)	(0.132)	(0.136)	(0.530)	(0.222)	(0.151)	(0.286)	(0.351)	(0.214)
Lagged acreage (A_{t-1})	0.124	0.419	0.625***	-0.525	1.048***	1.066***	0.594	0.373	0.772***	-0.332	1.152**	0.390**
	(0.965)	(0.311)	(0.165)	(1.438)	(0.169)	(0.169)	(0.646)	(0.328)	(0.173)	(0.239)	(0.476)	(0.205)
Co-efficient of multiple determination (R^2)	0.616	0.723	0.550	0.651	0.951	0.869	0.766	0.844	0.711	0.878	0.752	0.520
Co-efficient of adjustment (B)	0.876	0.581	0.375	1.525	-0.048	-0.066	0.406	0.627	0.228	1.332	-0.152	0.610
Short run acreage elasticity with respect to price	-0.4061	0.0793	0.0365	0.2690	-0.1878	0.0225	0.1335	0.0171	0.0147	-0.2884	0.1708	0.0009
Long run acreage elasticity with respect to price	-0.4636	0.1365	0.0973	0.1764	3.9125	-0.3409	0.3287	0.0273	0.0644	-0.2165	-1.1237	0.0015

Contd...

	Latur			*Parbhani*			*Nanded*			*Marathwada*		
	Sub-Period I	*Sub-period II*	*Overall period*	*Sub-period I*	*Sub-period II*	*Overall period*	*Sub-period I*	*Sub-period II*	*Overall period*	*Sub-period I*	*Sub-period II*	*Overall period*
	1980-1981 to 1989-1990	*1990-1991 to 1999-2000*	*1980-1981 to 1999-2000*	*1980-1981 to 1989-1990*	*1990-1991 to 1999-2000*	*1980-1981 to 1999-2000*	*1980-1981 to 1989-1990*	*1990-1991 to 1999-2000*	*1980-1981 to 1999-2000*	*1980-1981 to 1989-1990*	*1990-1991 to 1999-2000*	*1980-1981 to 1999-2000*
Constant in log	1.606	4.342	2.852	4.253	8.160	2.994	5.324	4.465	2.485	10.371	-3.155	4.093
terms (a)	(7.638)	(2.466)	(1.690)	(7.468)	(10.330)	(3.677)	(9.044)	(5.885)	(2.438)	(4.256)	(6.032)	(1.742)
Relative price	-0.904	-1.389	-1.047	0.220	-1.635	-0.292	0.176	-1.275	-0.054	-0.615	-0.751	0.135
lagged one year (RP_{t-1})	(2.927)	(1.417)	(0.727)	(0.771)	(1.407)	(0.668)	(1.185)	(1.131)	(0.828)	(0.582)	(0.349)	(0.322)
Relative yield	-0.262	0.695	0.141	0.296	-0.532	0.154	0.169	0.007	0.044	0.047	-0.556	0.082
lagged one year (RY_{t-1})	(0.600)	(0.690)	(0.238)	(0.278)	(0.703)	(0.285)	(0.333)	(0.315)	(0.167)	(0.107)	(0.404)	(0.117)
Gross irrigated	0.034	0.085	0.104	-1.135	-0.869	-0.569	-1.439	0.734	-0.363	-1.252	0.169	-0.401**
area lagged one year (GI_{t-1})	(0.793)	(0.295)	(0.181)	(1.301)	(1.824)	(0.640)	(2.129)	(0.959)	(0.356)	(0.889)	(0.558)	(0.202)
Sowing period	0.249	0.063	0.104	0.901*	-0.208	0.474*	1.112*	-1.101	0.370	0.913***	-0.034	0.250**
rainfall (RF_t)	(0.896)	(0.268)	(0.168)	(0.461)	(0.624)	(0.331)	(0.597)	(0.570)	(0.345)	(0.209)	(0.138)	(0.135)
Lagged acreage	0.572	0.433	0.480**	0.621	1.169*	0.711**	0.368	0.698**	0.526**	0.444	1.394**	0.698***
(A_{t-1})	(1.534)	(0.334)	(0.228)	(0.664)	(0.559)	(0.302)	(0.651)	(0.281)	(0.243)	(0.582)	(0.388)	(0.180)

Contd...

	Latur			*Parbhani*			*Nanded*			*Marathwada*		
	Sub-Period I	*Sub-period II*	*Overall period*	*Sub-period I*	*Sub-period II*	*Overall period*	*Sub-period I*	*Sub-period II*	*Overall period*	*Sub-period I*	*Sub-period II*	*Overall period*
	1980-1981 to 1989-1990	*1990-1991 to 1999-2000*	*1980-1981 to 1999-2000*	*1980-1981 to 1989-1990*	*1990-1991 to 1999-2000*	*1980-1981 to 1999-2000*	*1980-1981 to 1989-1990*	*1990-1991 to 1999-2000*	*1980-1981 to 1999-2000*	*1980-1981 to 1989-1990*	*1990-1991 to 1999-2000*	*1980-1981 to 1999-2000*
Co-efficient of multiple determination (R^2)	0.208	0.530	0.622	0.935	0.606	0.576	0.670	0.771	0.516	0.893	0.917	0.754
Co-efficient of adjustment (B)	0.428	0.567	0.520	0.379	-0.169	0.289	0.632	0.302	0.474	0.556	-0.394	0.302
Short run acreage elasticity with respect to price	-0.1563	-0.2925	-0.2025	0.0374	-0.3635	-0.0572	0.0369	-0.3182	-0.0124	-0.0939	-0.1353	0.0224
Long run acreage elasticity with respect to price	-0.3652	-0.5159	-0.3894	0.0986	2.1510	-0.1979	0.0584	-1.0536	-0.0262	-0.1689	0.3433	0.0742

Note: Coefficients for Jalna and Latur Districts are Calculated from 1981-82.
Figures in parentheses are standard errors.
*** - Significant at 1 per cent level
** - Significant at 5 per cent level
* - Significant at 10 per cent level

Sources: Computations are based on data collected from Season and Crop Reports, Epitome of Agriculture in Maharashtra, Part I & II.

of groundnut for sub-period II is found to be positive but insignificant in Aurangabad, Beed, Latur and Nanded districts. Coefficient of relative yield variable is positive and significant at 10 per cent level for Osmanabad district. It indicates the farmers tendency to shift their land area towards competing crop to get more yield. Coefficient of this variable is negative and insignificant in case of Jalna and Parbhani districts and Marathwada region as a whole. For overall period, the regression coefficients with respect to relative yield are found to be positive but not significant in case of Aurangabad, Beed, Osmanabad, Latur, Parbhani and Nanded districts and for Marathwada region as a whole and it is negative in case of Jalna district only.

The gross irrigated area lagged one year has exerted a negative bearing on acreage allocation decisions in Parbhani and Nanded districts and Marathwada region as a whole. It has exerted a positive bearing on acreage allocation decisions of the farmers in Aurangabad, Jalna, Beed, Osmanabad and Latur district and it is significant at 5 per cent level in Osmanabad district. It indicates that due to expansion of irrigation facilities, the area under groundnut has increased in Osmanabad district during the sub-period I. During the sub-period II, gross irrigated area lagged one year has exerted a negative bearing on acreage allocation decisions in Aurangabad, Beed, Osmanabad, Parbhani districts and positive impact found in case of Jalna, Latur and Nanded districts and Marathwada region as a whole. But for overall period, the gross irrigated area has shown a negative impact on acreage allocation decisions in Aurangabad, Jalna, Parbhani and Nanded districts and Marathwada region as a whole. The coefficient of this variable is negative and significant at 5 per cent level for Marathwada region. This explains that with the expansion of irrigation facilities, the area under groundnut has declined. Same result is found in the study conducted by Sable[9] in case of Nanded district. For the remaining districts, it is positive and insignificant.

The rainfall received during the sowing period has shown positive impact on acreage allocation decisions with respect to groundnut in case of Jalna, Latur, Parbhani, Nanded and Marathwada region as a whole for the sub-period I. The coefficient of rainfall variable is found to be significant at 10 per cent level in case of Parbhani district, at 10 per cent level in case of Nanded district and at 1 per cent level in case of Marathwada region for the same period. For Aurangabad, Beed and Osmanabad districts it is negative and insignificant. For the sub-period II, coefficients of rainfall variable are found to be positive in case of Jalna, Beed, Osmanabad and Latur districts and significant at 10 per cent level in case of Osmanabad district. The estimated parameters with respect to rainfall variable are negative and insignificant for Aurangabad, Parbhani and Nanded districts and for Marathwada region as a whole. But for the overall period, the coefficient with respect to rainfall is found to be significant at 5 per cent level in case of Marathwada region and at 10 per cent level in case of Parbhani district indicating that the favourable rains received during the sowing period would induce the farmers to employ more land to groundnut crop. Same result was found in the study conducted by Sable[10]. The impact of rainfall is found to be statistically insignificant in Aurangabad, Jalna, Beed, Osmanabad, Latur and Nanded districts for the same period.

The lagged acreage variable is one of the most important determinants of acreage variation for groundnut in all the districts and for Marathwada region as a whole. The impact of this variable is, however, found to be statistically significant at 1 per cent level in case of Aurangabad, Jalna, Beed and Marathwada region as a whole and at 5 per cent level in case of Osmanabad, Latur, Parbhani and Nanded districts for the overall period. The impact of this variable is found to be significant in case of most of the districts and Marathwada region as a whole for sub-period II also. This positive and significant bearing of lagged acreage variable suggests upward trend in area allocation to groundnut in all the districts and Marathwada region as a whole. Sable's[11] groundnut acreage response study for Marathwada region showed positive and significant co-efficient of lagged groundnut acreage variable.

The co-efficient of adjustment with respect to groundnut crop for Marathwada region is found to be 0.302 for overall period. It varies from -0.394 in sub-period II to 0.556 in sub-period I. The coefficient of adjustment with respect to groundnut crop for sub-period I is lowest in Parbhani district (0.379) and highest in Jalna district (1.525). For sub-period II it is highest in Beed district (0.627) and lowest in Parbhani district (-0.169). For overall period it is found to be lowest in Jalna district (-0.66) and highest in Osmababad district (0.610). The highest coefficient of adjustment with respect to groundnut in Jalna district indicates that the desired adjustments in acreage are more rapidly made and non-price factors rather than relative price factor operated more quickly. The lowest rate of adjustment in Parbhani district shows that the acreage under groundnut is influenced more by non-price factors and price factor operated very gradually and slowly. Among the non-price factors lagged acreage factor has been highly influential in Parbhani district.

The short run and long run acreage elasticities with respect to price indicates considerable variation among the districts of Marathwada region. The short run acreage elasticity for Marathwada region in overall period is 0.0224. But it is negative for sub-period I and II. The short run acreage elasticity is negative and very low (-0.4061) in case of Aurangabad district for the sub-period I, which indicates that only a small percentage of area under groundnut is affected by changes in price of groundnut in Aurangabad district. The highest short run acreage elasticity with respect to price is found in Jalna district (0.2690) for sub-period I, indicates that comparatively greater percentage of area under groundnut is affected by changes in price of groundnut. The long run acreage elasticity in Jalna district (3.9125) is higher for sub-period II, indicating that larger time is required for acreage adjustment in Jalna district in the same period.

Safflower

The estimated acreage response coefficients with respect to safflower for seven districts of Marathwada and for

Marathwada region as a whole along with the value of R^2, the coefficient of adjustment (B) and standard errors are given in the Table 2.

The value of R^2 for the overall period in Marathwada region is found to be 0.355 per cent, meaning that all the independent variables together explain only 36 per cent variation in the dependent variable and the remaining 64 per cent variation by the other variables not included in the equation. It means that the independent variables included in the equation did not play an important role in determining acreage of safflower crop in Marathwada region as a whole. The coefficient of determination was also very low for sub-period I (0.395) but it was very high (0.719) for sub-period II. The value of R^2 is found to be very low (0.392) for Parbhani district for overall period and it is high for Beed district (0.951) for sub-period II. It means that the independent variables play an important role in determining acreage of safflower in Beed district during this period.

For the overall period the regression coefficient of relative price variable is positive but not significant in case of Aurangabad, Jalna, Latur and Nanded districts and for Marathwada region as a whole. It is negative and insignificant for Beed, Osmanabad and Parbhani districts during the same period. But for sub-period II coefficient of relative price variable is positive and significant for Beed and Parbhani districts and for Marathwada region as a whole. It is significant at 5 per cent level in case of Beed, and Parbhani districts and at 10 per cent level for Marathwada region as a whole. The regression coefficient of relative price variable is positive and significant for Jalna and Latur districts. But these coefficients are negative and statistically insignificant in Aurangabad, Beed, Parbhani districts and positive but insignificant for Osmanabad and Nanded districts and for Marathwada region as a whole during the first sub-period.

The regression coefficients with respect to relative yield of safflower lagged one year is found to be positive and significant for Jalna, Beed, Osmanabad, Parbhani and Marathwada region as a whole for overall period. This coefficient is statistically

Table 2
Estimated Acreage Response Coefficients for Safflower in Marathwada Region

	Aurangabad			*Jalna*			*Beed*			*Osmanabad*		
	Sub-Period I	*Sub-period II*	*Overall period*	*Sub-period I*	*Sub-period II*	*Overall period*	*Sub-period I*	*Sub-period II*	*Overall period*	*Sub-period I*	*Sub-period II*	*Overall period*
	1980-1981 to 1989-1990	*1990-1991 to 1999-2000*	*1980-1981 to 1999-2000*	*1980-1981 to 1989-1990*	*1990-1991 to 1999-2000*	*1980-1981 to 1999-2000*	*1980-1981 to 1989-1990*	*1990-1991 to 1999-2000*	*1980-1981 to 1999-2000*	*1980-1981 to 1989-1990*	*1990-1991 to 1999-2000*	*1980-1981 to 1999-2000*
Constant in log	8.152	-5.019	3.523	4.539	0.509	7.149	-3.092	5.203	4.518	-4.307	9.917	-0.228
terms (a)	(3.078)	(13.395)	(2.359)	(1.975)	(4.231)	(2.564)	(2.602)	(1.494)	(2.559)	(4.250)	(4.815)	(2.771)
Relative price	-0.070	1.358	0.073	0.400**	-0.017	0.185	-0.477	0.451**	-0.185	0.378	-0.018	-0.573
lagged one year (RP_{t-1})	(0.451)	(1.220)	(0.462)	(0.169)	(1.108)	(0.444)	(0.180)	(0.141)	(0.234)	(0.697)	(0.453)	(0.380)
Relative yield	-0.344	0.024	0.028	0.361*	0.287	0.352**	-0.125	0.176**	0.152**	0.658**	0.158	0.324**
lagged one year (RY_{t-1})	(0.211)	(0.319)	(0.164)	(0.204)	(0.236)	(0.168)	(0.064)	(0.053)	(0.075)	(0.280)	(0.139)	(0.150)
Gross irrigated area	0.055	0.905	-0.324	0.316	0.847	-0.269	0.870**	0.201	-0.271	0.455	-0.517	0.482*
lagged one year (GI_{t-1})	(0.238)	(1.658)	(0.238)	(0.294)	(0.852)	(0.383)	(0.241)	(0.185)	(0.223)	(0.479)	(0.560)	(0.348)
Sowing period	0.002	0.355*	0.106	-0.070	0.145	0.061	0.125**	0.163***	0.087**	0.084	0.167**	0.049
rainfall (RF_t)	(0.092)	(0.204)	(0.099)	(0.078)	(0.202)	(0.109)	(0.048)	(0.027)	(0.048)	(0.069)	(0.054)	(0.038)

Contd...

	Aurangabad			*Jalna*			*Beed*			*Osmanabad*		
	Sub-Period I	*Sub-period II*	*Overall period*	*Sub-period I*	*Sub-period II*	*Overall period*	*Sub-period I*	*Sub-period II*	*Overall period*	*Sub-period I*	*Sub-period II*	*Overall period*
	1980-1981 to 1989-1990	*1990-1991 to 1999-2000*	*1980-1981 to 1999-2000*	*1980-1981 to 1989-1990*	*1990-1991 to 1999-2000*	*1980-1981 to 1999-2000*	*1980-1981 to 1989-1990*	*1990-1991 to 1999-2000*	*1980-1981 to 1999-2000*	*1980-1981 to 1989-1990*	*1990-1991 to 1999-2000*	*1980-1981 to 1999-2000*
Lagged acreage	-0.285	0.401	0.723***	-0.037	-0.139	0.088	0.463*	-0.374	0.547**	1.062*	-0.165	0.531**
(A_{t-1})	(0.579)	(0.360)	(0.192)	(0.042)	(0.880)	(0.171)	(0.290)	(0.154)	(0.224)	(0.553)	(0.266)	(0.245)
Co-efficient of multiple determination (R^2)	0.513	0.615	0.681	0.772	0.562	0.397	0.821	0.951	0.614	0.644	0.874	0.396
Co-efficient of adjustment (B)	1.285	0.640	0.277	1.037	1.139	0.912	0.537	1.374	0.453	-0.062	1.165	0.469
Short run acreage elasticity with respect to price	-0.0028	0.0326	0.0023	0.0566	-0.0024	0.0260	-0.0746	0.0754	-0.0299	0.0634	-0.0028	-0.0927
Long run acreage elasticity with respect to price	-0.0022	0.0509	0.0085	0.0546	-0.0021	0.0285	-0.1389	0.0549	-0.0660	-1.0226	-0.0024	-0.1977

Contd...

	Latur			Parbhani			Nanded			Marathwada		
	Sub-Period I	Sub-period II	Overall period	Sub-period I	Sub-period II	Overall period	Sub-period I	Sub-period II	Overall period	Sub-period I	Sub-period II	Overall period
	1980-1981 to 1989-1990	1990-1991 to 1999-2000	1980-1981 to 1999-2000	1980-1981 to 1989-1990	1990-1991 to 1999-2000	1980-1981 to 1999-2000	1980-1981 to 1989-1990	1990-1991 to 1999-2000	1980-1981 to 1999-2000	1980-1981 to 1989-1990	1990-1991 to 1999-2000	1980-1981 to 1999-2000
Constant in log terms (a)	6.824 (5.886)	8.781 (2.236)	5.851 (2.055)	-0.089 (2.938)	8.532 (5.085)	0.651 (2.802)	0.470 (14.776)	4.003 (2.968)	-0.667 (3.489)	-1.124 (6.014)	12.266 (8.014)	5.771 (3.162)
Relative price lagged one year (RP_{t-1})	1.490** (0.660)	-0.340 (0.655)	0.233 (0.450)	-0.652 (0.408)	5.502** (2.002)	-0.413 (0.453)	1.004 (2.788)	0.392 (0.371)	0.570 (0.642)	0.106 (0.448)	0.943* (0.587)	0.062 (0.337)
Relative yield lagged one year (RY_{t-1})	-0.162 (0.711)	0.576* (0.338)	-0.013 (0.265)	0.348* (0.190)	-0.388 (0.312)	0.314** (0.142)	0.134 (0.798)	0.514* (0.277)	0.214 (0.275)	-0.106 (0.306)	0.148 (0.216)	0.230* (0.165)
Gross irrigated area lagged one year (GI_{t-1})	-0.013 (0.538)	-0.551 (0.221)	-0.342 (0.265)	-0.127 (0.473)	-4.231 (1.754)	0.307 (0.402)	0.095 (2.413)	-0.480 (0.461)	0.216 (0.651)	0.419 (0.500)	-0.946 (1.154)	-0.038 (0.344)
Sowing period rainfall (RF_t)	-0.202 (0.175)	0.344** (0.148)	0.125 (0.105)	-0.033 (0.093)	-1.148** (0.364)	0.034 (0.090)	0.003 (0.292)	0.155 (0.130)	0.061 (0.101)	0.160 (0.184)	0.525** (0.215)	0.048 (0.095)
Lagged acreage (A_{t-1})	-0.243 (0.593)	-0.313 (0.283)	0.153 (0.236)	1.174** (0.463)	3.313** (1.088)	0.548* (0.365)	0.713 (0.512)	0.639 (0.333)	0.754*** (0.212)	0.603 (0.563)	0.189 (0.421)	0.277 (0.273)

Contd...

	Latur			Parbhani			Nanded			Marathwada		
	Sub-Period I	Sub-period II	Overall period	Sub-period I	Sub-period II	Overall period	Sub-period I	Sub-period II	Overall period	Sub-period I	Sub-period II	Overall period
	1980-1981 to 1989-1990	1990-1991 to 1999-2000	1980-1981 to 1999-2000	1980-1981 to 1989-1990	1990-1991 to 1999-2000	1980-1981 to 1999-2000	1980-1981 to 1989-1990	1990-1991 to 1999-2000	1980-1981 to 1999-2000	1980-1981 to 1989-1990	1990-1991 to 1999-2000	1980-1981 to 1999-2000
Co-efficient of multiple determination (R^2)	0.694	0.856	0.231	0.763	0.824	0.392	0.563	0.750	0.707	0.395	0.719	0.355
Co-efficient of adjustment (B)	1.243	1.313	0.847	-0.174	-2.313	0.452	0.287	0.361	0.246	0.397	0.811	0.723
Short run acreage elasticity with respect to price	0.0862	-0.0088	0.0095	-0.0315	0.1207	-0.0145	0.0108	0.0052	0.0069	0.0035	0.0146	0.0015
Long run acreage elasticity with respect to price	0.0693	-0.0067	0.0112	0.1810	-0.0522	-0.0321	0.0376	0.0144	0.0280	0.0088	0.0180	0.0021

Note: Coefficients for Jalna and Latur Districts are Calculated from 1981-82.
Figures in parentheses are standard errors.
*** - Significant at 1 per cent level
** - Significant at 5 per cent level
* - Significant at 10 per cent level

Sources: Computations are based on data collected from Season and Crop Reports, Epitome of Agriculture in Maharashtra, Part I & II.

significant at 5 per cent level in Jalna, Beed, and Osmanabad and Parbhani districts and at 10 per cent level in Marathwada region as a whole during the same period. For the first sub-period coefficient of relative yield variable is negative and insignificant in case of Aurangabad, Beed, Latur and Marathwada region as a whole. But it is positive and significant in Jalna at 10 per cent level, in Osmanabad at 5 per cent level and at 10 per cent level in Parbhani district. The coefficient of relative yield is statistically significant at 5 per cent level in Beed, at 10 per cent level in Latur and Nanded district for second sub-period. From this it is clear that there has been positive and significant influence of this variable on safflower acreage allocation decisions of farmers during all periods in Marathwada region.

The coefficient of gross irrigated area lagged one year for overall period is found to be negative and showing insignificant impact on acreage allocation decision in most of the districts and Marathwada region as a whole. But it is found positive and significant at 10 per cent level in Osmanabad district only for this period. Coefficient of gross irrigated area for sub-period I is positive and significant at 5 per cent level in case of Beed district only. For the second sub-period the coefficient of gross irrigated area is found to be statistically insignificant in all the districts of Marathwada and Marathwada region as a whole. It means irrigation facilities did not play an important role in acreage allocation decisions by farmers of Marathwada region.

The rainfall received during the sowing period has shown positive impact on acreage allocation decision with respect to safflower in Marathwada region for all the periods. But it is negative and insignificant for sub-period I in Jalna, Latur and Parbhani districts. The coefficient with respect to rainfall is found to be significant at 10 per cent level in case of Aurangabad district, at 1 per cent level in case of Beed district and at 5 per cent level in case of Osmanabad, Latur, Parbhani districts and Marathwada region as a whole during the second sub period. It is positive and statistically significant at 5 per cent level in Beed district only for overall period. Specially in case of Beed district the coefficient with respect to rainfall is found to be positive and significant for sub-period I, II and overall period indicating that the favourable rains received during the sowing period would induce the farmers to allocate more land to safflower crop in this district.

The impact of lagged acreage variable is found to be positive for safflower crop in all the districts and for Marathwada region as a whole for overall period. The impact of this variable is found to be statistically significant at 1 per cent level in Aurangabad and Nanded districts, at 5 per cent level in Beed and Osmanabad districts and at 10 per cent level in Parbhani districtan for the same period. This positive and significant impact of lagged acreage variable suggests upward trend in area allocation to safflower crop in the above districts. But for sub-period I the impact of this variable is found to be negative and insignificant in case of Aurangabad, Jalna and Latur districts and positive and significant at 10 per cent level in case of Beed and Osmanabad districts and at 5 per cent level in case of Parbhani district. The impact of this variable is, however, found to be statistically significant at 5 per cent level only in case of Parbhani district for the sub-period II. The coefficient of lagged acreage of safflower is found to be positive and statistically significant for all the periods in case of Parbhani district only. It means that the lagged acreage variable has emerged as one of the important determinants of safflower acreage allocation decisions of farmers in all the districts of Marathwada region.

The coefficient of adjustment with respect to safflower crop for Marathwada region is found to be 0.723 during the overall period. It varies from 0.397 for sub-period I to 0.811 for sub-period II. The coefficient of adjustment ranged from –2.313 in Parbhani district for sub-period II to 1.374 in Beed district for sub-period II. This highest coefficient of adjustment with respect to safflower indicates that the desired adjustments in acreage are more rapidly made and non-price factors rather than relative price factor operated more quickly in Beed district. The lowest rate of adjustment in Parbhani district shows that the acreage under safflower is influenced more by non-price factors and price factor operated very gradually and slowly in this district. Among the non-price factors relative yield, sowing period rainfall and lagged acreage factors have been highly influential in Parbhani district.

The short run acreage elasticity for Marathwada region during the overall period is 0.0015 but it is negative and very low in case of Osmanabad district (-0.0927) for this period. It indicates that only a small percentage of area under safflower is affected by changes in price of safflower in this district. The highest short run acreage elasticity with respect to price is found in Parbhani district (0.1207) during the second sub-period indicating comparatively higher percentage of area under safflower is affected by changes in price of safflower in this district.

The long run acreage elasticity is higher in Parbhani district for sub-period I, which indicates that the larger time is required for acreage adjustment.

Sunflower

The estimated acreage response coefficients with respect to sunflower for seven district of Marathwada region and Marathwada region as a whole along with the value of R^2, the coefficient of adjustment (B) and standard errors are given in the Table 3.

The value of R^2 is found to be 0.873, meaning that all the variables together explain 87 per cent variation in the dependent variable and remaining 13 per cent variation by other variables not included in the equation. It is clear that independent variables included in the equation are so much important in determining acreage of sunflower crop in Marathwada region as a whole during the overall period. The coefficient of multiple determination was higher (0.904) for sub-period I than that for sub-period II (0.348) in Marathwada region as a whole. The value of R^2 ranged from 0.140 for Nanded district during second sub-period to 0.971 for Aurangabad district during the first sub-period.

For overall period the regression coefficient of relative price variable is found to be positive but insignificant in Jalna,

Table 3
Estimated Acreage Response Coefficients for Sunflower in Marathwada Region

	Aurangabad			*Jalna*			*Beed*			*Osmanabad*		
	Sub-Period I	*Sub-period II*	*Overall period*	*Sub-period I*	*Sub-period II*	*Overall period*	*Sub-period I*	*Sub-period II*	*Overall period*	*Sub-period I*	*Sub-period II*	*Overall period*
	1980-1981 to 1989-1990	*1990-1991 to 1999-2000*	*1980-1981 to 1999-2000*	*1980-1981 to 1989-1990*	*1990-1991 to 1999-2000*	*1980-1981 to 1999-2000*	*1980-1981 to 1989-1990*	*1990-1991 to 1999-2000*	*1980-1981 to 1999-2000*	*1980-1981 to 1989-1990*	*1990-1991 to 1999-2000*	*1980-1981 to 1999-2000*
Constant in log	-0.101	7.101	3.888	-2.127	8.209	-0.563	-15.301	12.861	-3.408	2.877	13.879	2.488
terms (a)	(3.543)	(6.455)	(3.354)	(16.193)	(4.976)	(7.956)	(11.165)	(5.525)	(4.418)	(7.372)	(5.704)	(3.865)
Relative price	2.061*	0.637	0.839*	7.155**	1.311*	0.136	0.436	0.377	0.814*	1.582	0.328	0.254
lagged one year (RP_{t-1})	(1.118)	(0.465)	(0.592)	(2.830)	(0.700)	(0.878)	(0.900)	(0.506)	(0.542)	(1.629)	(0.446)	(0.642)
Relative yield	0.010	-0.488	-0.180	2.334**	-0.120	0.019	0.858*	0.061	0.332*	-0.189	0.131	0.009
lagged one year (RY_{t-1})	(0.317)	(0.251)	(0.267)	(0.843)	(0.195)	(0.387)	(0.404)	(0.152)	(0.202)	(0.473)	(0.099)	(0.166)
Gross irrigated	-1.020	-0.412	-0.594	-2.592	-0.895	0.074	2.445	-0.362	0.735	-1.235	-1.131	-0.686
area lagged one year (GI_{t-1})	(0.443)	(0.901)	(0.359)	(2.499)	(0.645)	(1.041)	(1.790)	(0.452)	(0.682)	(1.350)	(0.651)	(0.745)
Sowing period	1.445**	-0.340	0.080	3.623**	-0.139	0.206	0.149	-0.048	0.098	1.348*	0.064	0.590**
rainfall (RF_t)	(0.556)	(0.285)	(0.323)	(1.188)	(0.285)	(0.512)	(0.435)	(0.270)	(0.255)	(1.007)	(0.276)	(0.437)

Contd...

	Aurangabad			Jalna			Beed			Osmanabad		
	Sub-Period I	*Sub-period II*	*Overall period*	*Sub-period I*	*Sub-period II*	*Overall period*	*Sub-period I*	*Sub-period II*	*Overall period*	*Sub-period I*	*Sub-period II*	*Overall period*
	1980-1981 to 1989-1990	*1990-1991 to 1999-2000*	*1980-1981 to 1999-2000*	*1980-1981 to 1989-1990*	*1990-1991 to 1999-2000*	*1980-1981 to 1999-2000*	*1980-1981 to 1989-1990*	*1990-1991 to 1999-2000*	*1980-1981 to 1999-2000*	*1980-1981 to 1989-1990*	*1990-1991 to 1999-2000*	*1980-1981 to 1999-2000*
Lagged acreage (A_{t-1})	0.365** (0.160)	0.415* (0.270)	0.849*** (0.100)	0.356* (0.196)	0.874** (0.318)	0.792*** (0.127)	0.419* (0.260)	-0.545 (0.506)	0.599*** (0.137)	0.799** (0.318)	0.096 (0.401)	0.880*** (0.179)
Co-efficient of (R^2) multiple determination	0.971	0.650	0.943	0.942	0.754	0.843	0.938	0.362	0.887	0.741	0.578	0.721
Co-efficient of adjustment (B)	0.635	0.585	0.151	0.644	0.126	0.208	0.581	1.545	0.401	0.201	0.904	0.120
Short run acreage elasticity with respect to price	0.3581	0.1071	0.1427	-0.0511	-0.9303	-0.0064	-0.0222	-0.0274	-0.0237	-0.0712	-0.0250	-0.0156
Long run acreage elasticity with respect to price	0.5639	0.1830	0.9450	-0.0793	-7.3833	-0.0308	-0.0382	-0.1770	-0.0591	-0.3543	-0.0276	-0.1300

Contd...

	Latur			Parbhani			Nanded			Marathwada		
	Sub-Period I	*Sub-period II*	*Overall period*	*Sub-period I*	*Sub-period II*	*Overall period*	*Sub-period I*	*Sub-period II*	*Overall period*	*Sub-period I*	*Sub-period II*	*Overall period*
	1980-1981 to 1989-1990	*1990-1991 to 1999-2000*	*1980-1981 to 1999-2000*	*1980-1981 to 1989-1990*	*1990-1991 to 1999-2000*	*1980-1981 to 1999-2000*	*1980-1981 to 1989-1990*	*1990-1991 to 1999-2000*	*1980-1981 to 1999-2000*	*1980-1981 to 1989-1990*	*1990-1991 to 1999-2000*	*1980-1981 to 1999-2000*
Constant in log terms (a)	0.778 (2.970)	4.377 (4.059)	2.268 (1.417)	-11.743 (19.278)	2.675 (8.326)	-0.287 (12.399)	1.152 (12.609)	5.683 (8.151)	3.215 (6.731)	3.993 (12.332)	12.826 (7.272)	3.467 (5.579)
Relative price lagged one year (RP_{t-1})	-0.367 (3.014)	0.157 (0.444)	0.171 (0.236)	7.681* (4.540)	0.245 (0.749)	0.439 (1.405)	-3.281 (2.495)	-0.007 (0.527)	-0.472 (0.841)	2.530* (1.625)	0.109 (0.437)	0.225 (0.523)
Relative yield lagged one year (RY_{t-1})	0.066 (0.483)	0.240 (0.479)	0.241 (0.208)	-0.582 (1.891)	-0.076 (0.266)	-0.116 (0.555)	-1.006 (0.965)	0.033 (0.409)	-0.146 (0.430)	0.093 (0.385)	0.195 (0.167)	0.058 (0.164)
Gross irrigated area lagged one year (GI_{t-1})	0.178 (0.792)	-0.496 (0.483)	-0.423 (0.256)	0.514 (2.872)	-0.184 (0.995)	-0.111 (1.743)	1.061 (2.848)	0.924 (2.125)	-0.333 (1.030)	-0.692 (1.795)	-0.404 (0.821)	-0.332 (0.753)
Sowing period rainfall (RF_t)	-0.084 (1.024)	0.137 (0.449)	0.113 (0.212)	1.858 (1.541)	0.159 (0.543)	0.317 (0.950)	-0.959 (1.664)	-0.360 (0.797)	-0.044 (0.609)	1.075 (0.810)	-0.324 (0.360)	0.110 (0.293)
Lagged acreage (A_{t-1})	0.816* (0.410)	0.681 (0.766)	0.948*** (0.180)	0.172 (0.397)	0.632* (0.387)	0.880*** (0.177)	0.660* (0.376)	-0.734 (1.307)	0.864* (0.183)	0.480* (0.295)	0.087 (0.492)	0.870*** (0.137)

Contd...

	Latur			*Parbhani*			*Nanded*			*Marathwada*		
	Sub-Period I	*Sub-period II*	*Overall period*	*Sub-period I*	*Sub-period II*	*Overall period*	*Sub-period I*	*Sub-period II*	*Overall period*	*Sub-period I*	*Sub-period II*	*Overall period*
	1980-1981 to 1989-1990	*1990-1991 to 1999-2000*	*1980-1981 to 1999-2000*	*1980-1981 to 1989-1990*	*1990-1991 to 1999-2000*	*1980-1981 to 1999-2000*	*1980-1981 to 1989-1990*	*1990-1991 to 1999-2000*	*1980-1981 to 1999-2000*	*1980-1981 to 1989-1990*	*1990-1991 to 1999-2000*	*1980-1981 to 1999-2000*
Co-efficient of multiple determination (R^2)	0.861	0.285	0.792	0.878	0.455	0.811	0.839	0.140	0.835	0.904	0.348	0.873
Co-efficient of adjustment (B)	0.184	0.319	0.052	0.828	0.368	0.120	0.340	1.734	0.136	0.520	0.913	0.130
Short run acreage elasticity with respect to price	-0.0191	0.0007	0.0045	-0.0048	-0.0157	-0.0178	-0.1189	0.0075	-0.0066	-0.0484	-0.0057	-0.0082
Long run acreage elasticity with respect to price	-0.1038	0.0022	0.0865	-0.0059	-0.0426	-0.1483	-0.3497	-0.0043	-0.0485	-0.0931	-0.0062	-0.0631

Note: Coefficients for Jalna and Latur Districts are Calculated from 1981-82.
Figures in parentheses are standard errors.
*** - Significant at 1 per cent level
** - Significant at 5 per cent level
* - Significant at 10 per cent level

Sources: Computations are based on data collected from Season and Crop Reports, Epitome of Agriculture in Maharashtra, Part I & II.

Osmanabad, Latur and Parbhani districts and Marathwada region as a whole. For this period coefficient of price variable is positive and statistically significant at 10 per cent level in Aurangabad and Beed districts. This coefficient is negative and insignificant for all the periods in Nanded district. Coefficient of relative price variable for sub-period one is found to be positive and significant in Aurangabad at 10 per cent level, in Jalna at 5 per cent level and at 10 per cent level in Marathwada region as a whole. But it is negative and insignificant for Latur and Nanded districts for the same period. For sub-period II, the coefficient of relative price variable is statistically significant at 10 per cent level in Parbhani district only. Positive and significant value of this variable shows that relative price of sunflower lagged one year has significant impact on sunflower acreage allocation decisions of the farmers in most of the districts of the region.

The regression coefficients with respect to relative yield variable are found to be positive, but not significant for sub-period I, in case of Aurangabad, Beed and Latur districts and Marathwada region as a whole. But it is positive and significant at 5 per cent level in Jalna district and at 10 per cent level in Beed district and negative and insignificant for Osmanabad, Parbhani and Nanded districts for sub-period I. Coefficients with respect to relative yield of sunflower during the sub-period II are found to be negative and insignificant in Aurangabad, Jalna, Parbhani and Nanded districts and positive but insignificant in Beed, Osmanabad, and Latur districts. For overall period it is positive but not significant in Jalna, Osmanabad and Latur districts and Marathwada region as a whole. But it is positive and significant at 10 per cent level in Beed district only and negative and insignificant in Aurangabad, Parbhani and Nanded districts for this period. It is clear that for all the periods the coefficient of relative yield of sunflower lagged one year has shown positive but insignificant impact on sunflower acreage allocation decisions of farmers in Marathwada region as a whole.

The coefficient of gross irrigated area has shown negative and insignificant impact on acreage allocation decisions in all the districts of Marathwada region for all the periods. This negative value indicates that the irrigation facilities, which are in existence are not important in determining acreage allocation decisions of farmers in Marathwada region. Exceptionally the coefficient of gross irrigated area lagged one year is found to be positive but insignificant in Jalna and Beed districts for overall period, as well as for sub-period one in Beed district also.

For overall period the coefficient with respect to rainfall variable is found to be positive but insignificant in Aurangabad, Jalna, Beed, Latur and Parbhani districts and Marathwada region as a whole. But for this period it is positive and statistically significant at 5 per cent level in Osmanabad district and negative and insignificant in case of Nanded district. The coefficient with respect to rainfall is found to be significant at 5 per cent level in Aurangabad and Jalna districts and for Osmanabad district it is significant at 10 per cent level and for Parbhani district and Marathwada region as a whole it is found to be positive but insignificant and negative and insignificant in case of Latur and Nanded districts for the sub-period one. For sub-period II the coefficient with respect to rainfall is found to be negative and insignificant in case of Nanded district and Marathwada region as a whole. The rainfall received during the sowing period has shown positive impact on acreage allocation decisions with respect to sunflower in case of Marathwada region as a whole for the overall period. It means that favourable rainfall during sowing period would induce the farmers to allocate more land to sunflower crop.

The lagged acreage variable is one of the most important determinants of acreage of sunflower in all the districts and for Marathwada region as a whole. The impact of this variable is found to be positive and significant in all the districts and for Marathwada region as a whole for all the periods of time. For over all period the impact of this variable is found to be statistically significant at 1 per cent level in Aurangabad, Jalna, Beed, Osmanabad, Latur, Parbhani districts and for Marathwada region as a whole and significant at 10 per cent level in case of Nanded district during this period. The impact of this variable during the sub-period first is found to be significant at 10 per cent level in case of Jalna, Beed, Latur and Nanded districts and for Marathwada region as a whole. It is significant at 5 per cent level in Aurangabad and Osmanabad districts, and positive but insignificant in Parbhani district for this sub-period. For the sub-period II, the impact of this variable is found to be statistically significant at 10 per cent level in Aurangabad and Parbhani districts. It is significant at 5 per cent level in case of Jalna district. But in case of Beed and Nanded districts it is found negative and insignificant and in case of Osmanabad and Latur districts and Marathwada region as a whole it is found to be positive but insignificant for sub-period II. Positive and significant bearing of lagged acreage variable shows upward trend in area allocation to sunflower in all the districts and for Marathwada region as a whole during the overall period.

The coefficient of adjustment with respect to sunflower crop in Marathwada region as a whole varies from 0.520 for sub-period I to 0.913 for sub-period II and for overall period it is 0.130. The lowest rate of adjustment shows that the acreage under sunflower is influenced by non-price factors and price factor operated very slowly in the region as a whole. The district level picture shows that the coefficient of adjustment with respect to sunflower for overall period ranged from 0.052 in Latur district to 0.401 in Beed district. The highest coefficient of adjustment with respect to sunflower in Beed district indicates that the desired adjustments in acreage of sunflower in Beed district are made rapidly and non-price factors rather than relative price factor operated more quickly. The lowest rate of adjustment in Latur district shows that the acreage under sunflower in this district is influenced by non-price factors and price factor operated slowly. It means that only lagged acreage variable is important among the non-price factors in Latur district. The coefficient of adjustment with respect to sunflower crop varies from 0.126 for sub-period II in Jalna district to 1.734 for sub-period II in Nanded district.

The short run acreage elasticity with respect to price in overall period for Marathwada region as a whole is -0.0082.

But it is positive and very high in case of Aurangabad district (0.1427) indicating a comparatively greater percentage of area under sunflower is affected by changes in price of sunflower in this district during the overall period. But it is negative (-0.0237) and very low in case of Beed district, which indicates that only a small percentage of area under sunflower is affected by changes in price of sunflower in this district. The long run acreage elasticity with respect to price for overall period in Marathwada region as a whole is negative (-0.0631) and very low indicating that smaller time is required for acreage adjustment in the region as a whole. Long run acreage elasticity with respect to price for Aurangabad district is higher (0.9450) indicating that larger time is required for acreage adjustment during the overall period in Aurangabad district

Soyabean

The estimated acreage response coefficients with respect to soyabean for Marathwada region along with the value of R^2, the coefficient of adjustment (B) and standard errors are given in Table 4.

Table 4

Estimated Acreage Response Coefficients for Soyabean in Marathwada Region

	1990-91 to 1999-2000
Constant in log terms (a)	1.852(26.969)
Relative price lagged one year (RP_{t-1})	0.660(1.193)
Relative yield lagged one year (RY_{t-1})	-0.064(0.486)
Gross irrigated area lagged one year (GI_{t-1})	-0.211(3.008)
Sowing period rainfall (RF_t)	0.361(0.722)
Lagged acreage (A_{t-1})	0.706**(0.241)
Co-efficient of multiple determination (R^2)	0.796
Co-efficient of adjustment (B)	0.294
Short run acreage elasticity with respect to price	-0.0340
Long run acreage elasticity with respect to price	-0.1156

Note: *** - Significant at 1 percent level
** - Significant at 5 per cent level
* - Significant at 10 per cent level
Figures in parentheses are standard errors.

Sources: Computations are based on data collected from Season and Crop Reports and Epitome of Agriculture in Maharashtra, Part I & II.

Five independent variables are used to identify their impact on soyabean acreage during the period 1990-91 to 1999-2000 in Marathwada region. Relative price of soyabean lagged one year, relative yield of soyabean lagged one year, gross irrigated area lagged one year, sowing period rainfall and lagged acreage of soyabean are the explationary variables, which are incorporated in the equation. For the period 1990-91 to 1999-2000, the value of R^2 was found to be 0.796 per cent, meaning that all the variables together explain about 80 per cent variation in dependent variable and remaining 20 per cent variation by the other variables not included in the equation. It is clear that the independent variables included in the equation play an important role in determination of acreage of soyabean crop in Marathwada region.

The regression coefficient of relative price of soyabean lagged one year is found to be positive but insignificant for Marathwada region. This shows that the relative price of soyabean lagged one year has shown insignificant impact on soyabean acreage allocations of farmers in Marathwada region.

The regression coefficient of relative yield of soyabean lagged one year is found to be negative and insignificant in Marathwada region. The coefficient of relative yield has shown negative and insignificant impact on soyabean acreage allocation decisions of farmers in Marathwada.

The regression coefficient of gross irrigated area lagged one year is found to be negative and insignificant in Marathwada region. This value indicates that the irrigation facilities are not important in determining acreage allocation decisions of farmers in Marathwada region. The rainfall received during the sowing period has shown positive and insignificant impact on acreage allocation decisions of farmers in Marathwada region.

The lagged acreage variable has emerged as one of the important determinants of acreage variation for soyabean in Marathwada region as a whole. The regression coefficient of one year lagged acreage of soyabean exerted positive impact on acreage allocation decisions of farmers in Marathwada region. The coefficient of lagged acreage is found to be significant at 5 per cent level for the period 1990-91 to 1999-2000 in Marathwada region.

The coefficient of adjustment with respect to soyabean crop for the period 1990-91 to 1999-2000 is found to be 0.294 for Marathwada region. This lowest rate of adjustment shows that the acreage under soyabean is influenced more by non-price factors and price factor operated very gradually and slowly. Among the non-price factors the lagged acreage factor has been highly influential in Marathwada region.

The short run acreage elasticity with respect to price is -0.0340 and long run acreage elasticity with respect to price is -0.1156 in Marathwada region. The long run elasticity with respect to price is negative and very low, which indicates that only a small percentage of area under soyabean is affected by changes in price of soyabean in Marathwada region.

CONCLUSIONS

The above analysis leads to the following conclusions:

The impact of relative price variable with respect to groundnut is found to be positive , but insignificant in Marathwada region as a whole during the overall period. The district level pictures reveal that there is positive impact of price variable in four districts and negative impact in three districts. But it is insignificant in all the districts of the region. The relative yield factor is found to be positive but insignificant in six districts and Marathwada region as a whole during overall period and negative and insignificant in Jalna district only. Yield variable has shown positive and significant impact in Osmanabad district during sub-period II. The gross irrigated area has indicated negative impact on groundnut acreage in most of the districts of the region but in case of Marathwada region as a whole it indicated negative and significant impact during the overall period. Impact of this factor is positive and significant in case of Osmanabad district during sub-period one. The sowing period rainfall has positively and significantly affected the area under groundnut in all the districts and Marathwada region as a whole during the overall period.

The lagged acreage variable has positive and significant impact on acreage under groundnut in all the districts of Marathwada and Marathwada region as a whole. The coefficient of acreage adjustment for groundnut is found to be highest (0.520) in Latur district and the lowest (-0.066) in Jalna district during the overall period of time. The value of R^2 is 0.754 for overall period indicating that independent variables used in equation explain 75 per cent variation in dependent variable. The coefficient of adjustment with respect to groundnut crop for Marathwada region is found to be 0.302 for overall period. This lowest rate of adjustment indicates that the acreage under groundnut is influenced more by non-price factors and price factor operated very slowly in Marathwada region. The short run acreage elasticity with respect to groundnut price for Marathwada region in overall period is 0.0224. This lowest value of elasticity shows smaller percentage of area under groundnut is affected by changes in price of groundnut during overall period in Marathwada. The long run acreage elasticity with respect to price is 0.0742. This lowest value indicates that larger time is not required for acreage adjustment in Marathwada region.

The sunflower producers have positively, but not significantly responded to relative price of sunflower in Jalna, Osmanabad, Latur and Parbhani districts and in Marathwada region as a whole. The impact of yield variable in case of sunflower is found to be positive though statistically insignificant in Jalna, Osmanabad, Latur and Nanded districts and in Marathwada region but negative and insignificant in Aurangabad and Parbhani districts and significant at 5 per cent level in Beed district during the overall period. Gross irrigated area lagged one year has indicated negative impact on sunflower acreage in four districts and in Marathwada region except Jalna, Beed and Nanded districts during the overall period. The sowing period rainfall has significantly affected the area under sunflower in case of Osmanabad district. In the remaining cases and for Marathwada region as a whole, the influence of rainfall is insignificant during the overall period. Lagged acreage variable has positive and highly significant influence on acreage under sunflower in all the districts and in Marathwada region as a whole, which means that lagged acreage variable suggests upward trend in area allocation to safflower in Marathwada region.

The value of coefficient of multiple determination is found to be 0.873 for Marathwada region as a whole. This value indicates that the independent variables respond to variation in sunflower acreage in Marathwada region. The value of R^2 is highest in Aurangabad district (0.943) and lowest in Osmanabad district (0.721). The highest value of R^2 indicate that the independent variables play an important role in determining sunflower acreage in Aurangabad district during the overall period.

The coefficient of adjustment with respect to sunflower is found to be 0.130 for Marathwada region as a whole. The district level picture reveals that the coefficient of adjustment is found highest in Beed district (0.401) and lowest in Latur district (0.052) in the overall period. Highest rate of adjustment with respect to sunflower in Beed district indicates that the desired adjustments in sunflower acreage are more rapidly made and non-price factors rather than relative price factor operated more quickly in Beed district during the overall period. For the overall period, the short run acreage elasticity with respect to sunflower price in Marathwada region is -0.0082 and long run acreage elasticity is –0.0631. This negative and smaller value shows that smaller time is required for sunflower acreage adjustment in Marathwada region as a whole during the overall period of time.

Relative price of safflower lagged one year is found to be insignificant in Marathwada region as a whole during the overall period. The relative yield variable has exerted positive influence and it is significant at 10 per cent level in Marathwada region as a whole as well as in Beed, Osmanabad and Parbhani districts. It is negative and insignificant in Latur district during the overall period. The gross irrigated area has indicated positive impact on safflower acreage in Osmanabad, Parbhani and Nanded districts and is significant at 10 per cent level in Osmanabad district but negative and insignificant in case of

Marathwada region as a whole during the overall period. The rainfall received during sowing period has positively affected the area under safflower in all the districts and Marathwada region as a whole and is significant at 5 per cent level in Beed district only during the overall period. Lagged safflower acreage has positive and highly significant impact on acreage in Aurangabad,. Beed, Osmanabad, Parbhani and Nanded districts. It is the most important factor determining the acreage under safflower in most of the districts of Marathwada and only positive in Marathwada region as a whole during the overall period. This positive and significant impact of lagged acreage variable suggests upward trend in area allocation to safflower crop in all districts and in the region also.

The coefficient of adjustment for safflower is found to be 0.723 during the overall period in Marathwada region as a whole and it is found highest in Jalna district (0.912) and lowest in Nanded district (0.246) for the same period. The lowest rate of adjustment with respect to safflower in Nanded district indicates that the acreage under safflower is influenced more by non-price factors and price factor operated very slowly in this district during the overall period.

The short run acreage elasticity with respect to safflower price for Marathwada region as a whole during the overall period is 0.0015. This smaller value shows smaller percentage of area under safflower is affected by changes in price of safflower during the overall period in Marathwada region as a whole. The long run acreage elasticity with respect to safflower price is 0.0021 for Marathwada region during the overall period. This smaller value shows that larger time is required for acreage adjustment in Marathwada region during the same period.

Relative price variable has positively affected acreage allocation decisions of the farmers in case of soyabean in Marathwada region. The impact of relative yield is found to be negative and insignificant. The gross irrigated area has indicated negative impact on soyabean acreage in Marathwada region. The rainfall received during sowing period has positively affected the area under soyabean in this region.

Lagged acreage variable also exerted positive and significant influence on acreage of soyabean crop. The coefficient of adjustment for soyabean is found to be 0.294. This lowest rate of adjustment with respect to soyabean indicates that the acreage under soyabean is affected more by non-price factors and relative price factor operated very gradually in Marathwada region. The short run acreage elasticity with respect to price is -0.0340 and long run elasticity with respect to price is -0.1156, means that long run acreage elasticity is smaller, that means smaller time is required for acreage adjustment in Marathwada region during the overall period larger.

NOTES AND REFERENCES

1. Reddy, J. M. (1970), Estimation of Farmers Supply Response–A Case Study of Groundnut, *Indian Journal of Agricultural Economics*, Vol. XXV, No. 4.
2. Jhala, M. L. (1979), Farmers Response to Economic Incentives-An Analysis of Interregional Groundnut Supply Response in India, *Indian Journal of Agricultural Economics*, Vol. XXIV, No. 1.
3. Kaul, J. L. and Sidhu, D. S. (1971), Acreage Response to Prices for Major Crops in Punjab–An Econometric Study, *Indian Journal of Agricultural Economics*, Vol. XXVI, No. 4.
4. Madhavan, M. C. (1972), Acreage Response of Indian Farmers-A Case Study of Tamil Nadu, *Indian Journal of Agricultural Economics*, Vol. XXVII, No. 1.
5. Sable, M. S. (1991), Supply Response of Some Agricultural Commodities in Marathwada, Unpublished Ph.D. Thesis, Marathwada University, Aurangabad, August.
6. Jhala, M. L., *Op.cit*.
7. Sable, M. S., *Op.cit*.
8. Reddy, J. M., *Op.cit*.
9. Sable, M. S., *Op.cit*.
10. Sable, M. S., *Op.cit*.
11. Sable, M. S., *Op.cit*.

6 Findings and Policy Measures

This is the final chapter. It attempts to give summary of the earlier chapters, to recapitulate the important findings of the study and mention some policy measures.

India is a developing country where the population has been expanding very rapidly and the levels of income are rising. Indian economy is an agricultural economy wherein near about 70 per cent population is dependent on agriculture. Agriculture sector plays an important role in the economic development of the economy by providing food and raw materials, employment, capital and surpluses for national economic development. It creates demand for industrial consumer goods. It feeds to people and cattle by providing food and fodder. It provides raw materials like cotton, jute, sugarcane, oilseeds to manufacturing industries. Oilseeds are one of the age old products of this earth and they play an important role in the human diet. India is one of the most leading oilseeds producing country of the world. Oilseeds play an important role in the economic development of the country and it is an important base for the agro-industries such as oil milling industry, soap industry, paint industry, hair oil industry and toiletries industry etc.

Agriculture is the fundamental basis of overall economic development and has the potential of becoming a leading sector in the process of economic development. So efforts are to be made to increase the production of agriculture. To increase the agriculture production one should know the important factors to which it responds more. Agriculture production is a complex process in which price and non-price factors play significant role. The. price expectations in a commercialised agriculture play an eminent role in decision making at micro and macro levels. The price is one of the most important factors affecting agricultural production on one hand and agricultural economy on the other hand. The farmers make an appropriate adjustments in acreage under crops in response to changes in the prices. The development of agriculture requires a sound and healthy agricultural policy. Agricultural price policy is one of the important instruments in the hands of the government to achieve higher growth rates in agricultural production. In order to make price policy an effective instrument for inducing desired changes in the resource allocation to different crop enterprises, the knowledge of farmers acreage responsiveness is essential. Two opposite views are observed regarding the responsiveness of agriculture to price factors. One view held by Dalton and other agricultural economists is that the farmers in underdeveloped countries required only a little amount of cash. Mathur and Ezekeil opined that farmers respond negatively to changes in price as they can get the required amount of cash by selling less and keep more for their own consumption. According to Rajkrishna, the farmers respond favourably to price changes, therefore, prices are the most important determinants of the area under different crops. Therefore, agriculture price policy should be positive. However, Dharam Narain, Narayanan and others have asserted that agriculture is responsive to price changes.

It is essential to develop policies in order to stop deterioration in the commodity terms of trade and reduction in the earning of agricultural exports, deficits in the balance of payments, unemployment, in framing policies related to government price support programmes, tariffs and the knowledge of supply response to price changes is must.

With the introduction of high yielding varieties of wheat after mid-sixties and of rice in early seventies these two crops became markedly more profitable to grow compared with other crops in the area where assured irrigation facilities were available. This continuous change spread over to decades solved the food problem of the country in respect to cereals, but inter alia, diversion of irrigated crop lands at the margin from under oilseeds did not permit the growth of oilseeds production to match the domestic demand for edible oils. Although India is one of the largest producers of oilseeds, the edible oils production falls short of the country's requirements. This has necessitated large scale imports of edible oils. About 10 per cent of the total cropped area in the country is estimated to be under oilseed crops, the more important of which the groundnuts, rapeseed, mustard, sesamum, linseed, castor and soyabean. About 92 per cent of the total land under oilseeds crops is rainfed and hence the crop prospects are uncertain.

Economists have extensively investigated the growth performance of rice and wheat during the last four decades. It has been widely researched at the global, national, state level, regional and household levels. Unfortunately, scant attention has been paid to the study of oilseed crops, which play an important role in sustaining crop system and meeting the domestic needs of the edible oil in the country. Therefore, an attempt must be made to provide current evidence on the temporal and spatial dimensions of the oilseed development. The production of oilseeds like other agricultural commodities is primarily determined by acreage. Therefore, it would be worthwhile to investigate the factors influencing acreage. Various price and non-price factors influence the farmer's decisions regarding land allocation to different crop. The first segment includes input and output prices. These range from last year's harvest prices of the crops, availability of minimum support price, last year's harvest price of the competing crop to prices of fertilizers, power, seed, water, insecticides and availability of credit. Similarly, a host of non-price factors also play an important role. The major factors are last year's acreage and yield, availability of improved seeds and irrigation, rainfall, facility of procurement by government agencies, resistance of crops to pest attacks, extension services, home consumption, availability of alternate crops, credit and assured market. Unfortunately, crop specific information on all these variables is lacking and therefore, even tabular analysis is not possible. Owing to this difficulty, the findings of acreage response studies are based on a few variables for which data are available. Thus, factors affecting oilseed acreage are numerous and their contribution is generally estimated through an acreage response model. Most of the studies cover a period upto eighties but policy scenario has dramatically changed after gradual liberalization of agriculture. Moreover, the nature of acreage response differs from crop to crop and region to region.

Marathwada is the most backward region of Maharashtra state and the economy of Marathwada is predominantly an agricultural economy. It is endowed with natural resources such as rainfall and soils are relatively more favourable in this region. The soil and climate of this region are favourable to take oilseed crops such as groundnut, safflower, sunflower and soyabean. Marathwada shares about 12.5 per cent of the states total area under oilseeds and around 11 per cent of state's total oilseed production. There is more scope in this region to increase area under oilseeds. The crop gain or loss area depending on their relative profitability and the farmer's resource for exploiting those possibilities. The acreage allocation decisions of farmers are always made by comparing relative returns. Further rainfall, irrigation and some other factors also cause the acreage allocation decisions of the farmers. It will, therefore, be interesting to probe into the factors, which determine the acreage under major oilseeds in Marathwada. No such study has so far been conducted for oilseed crops in this region for the eighties and nineties. This study aims to study the growth in oilseeds production and factors which are important in making decisions regarding allocation of area to oilseed crops in Marathwada region as a whole and district level for the period 1980-81 to 1999-2000. Main objectives of the study were:

1. To study the growth in area, production and yield of oilseed crops in Marathwada;
2. To study the area response of important oilseed crops in Marathwada;

3. To probe into the factors which are important in determining acreage of oilseed crops in Marathwada.
4. To suggest policy measures to increase acreage and production of oilseeds in Marathwada region.

This study was based on secondary data. The required data on area, production and yield and prices of important oilseed crops were collected from various government publications such as Season and Crop Reports, Epitome of Agriculture in Maharashtra, Part I and II, District Statistical Abstracts, Population Census of India and Farm Harvest Prices in India. The study period was divided into two sub-periods. The first sub-period relates to 1980-81 to 1989-90 and the second sub-period relates to 1990-91 to 1999-2000. These represent the pre- and post-reforms periods. Hingoli district was clubbed with Parbhani district. It covered four important oilseeds, namely, groundnut, safflower, sunflower and soyabean.

To calculate growth rates of area, production and yield of major oilseed crops a semi-log function of the form $Y = AB^x$ or $\log Y = \log A + x. \log B$ was used and to examine the acreage response of major oilseed crops Nerlovian adjustment lag model was used. The reduced form of equation was, $A_t = a + b_1 A_{t-1} + b_2RP_{t-1} + b_3RY_{t-1} + b_4GI_{t-1} + b_5RF_t + U_t$.

To examine the acreage response of major oilseed crops during the current year, oilseed acreage was regressed on lagged acreage, lagged relative price, lagged relative yield, lagged gross irrigated area and sowing period rainfall in the different districts of Marathwada region. One year lag was used in acreage, yield, price and irrigation assuming that the current acreage, yield, irrigation and price generally influence the decisions about area allocation in the next year.

Farmer is a central point who makes final decisions regarding the allocation of land to different crops. It is generally hypothesized that an acreage in the price induces the farmers to increase the area under cultivation to the best possible extent, to apply more inputs to produce the maximum possible quantum using the best available technology. Generally two different approaches are used to measure empirically the supply or acreage responsiveness: (1) the constructive approach and (2) statistical analysis of time series data in which three methods viz. graphical or tabular method, regression or correlation coefficient analysis and multiple regression analysis are included. The multiple regression analysis method is further subdivided into two groups: (i) traditional models where adjustment mechanism is not specified, and (ii) adjustment lag models. In almost all the studies on acreage response, the acreage response has been examined by making use of statistical analysis of time series in which multiple regression analysis and again Nerlovian adjustment lag model has been used by many researchers.

A considerable amount of literature has appeared on acreage response. At present two different views on acreage response prevail. One view is that the farmers in underdeveloped countries are not responsive to the changes in relative price and the other view is that they are no less responsive than their counterparts of the developed countries like the United States. Marc Nerlove, an eminent economist, found that the farmers in the United States are responsive to changes in relative prices. Mishra and Sinha's study reveals that in a country like India, in case of commercial crops higher relative prices may increase in acreage of crops. Raj Krishna's study reveals that one year lagged post-harvest price was important in determining acreage in Punjab. John's study indicated that in the majority of cases positive relationship of one kind or another exist between price and area/ output or relative price and relative area/output. Kamala Devi and Rajgopalan's study represents that an increase in the relative price influences the acreage under groundnut in the following year in North Arcot district of Madras state. Dharam Narayan in his study relating to the impact of price movements on areas under selected crops in India has found that the Indian farmers are significantly responsive to price. Rajgopalan found stronger and more consistent evidence for a significant role of prices in influencing groundnut output than it did for any of the other crops he considered in Madras state. Behrman in his study relating to supply response in case of four major annual crops in Thailand has found that large variations in prices had an adverse impact on acreage.

Agriculture is one of the most important sector of the Marathwada region. In the year 2001 about 76.72 per cent of the workers in this region were engaged in agriculture. According to 1992-93 data, about 78 per cent of the workers were engaged in agriculture which reveals that the major portion of the population of this region depend on agriculture.

Ecological and geo-chemical atmosphere is favourable to agriculture in Marathwada region. Godavari, Purna and Manjra are the important rivers of this region. The soils vary greatly in texture and depth. In the years of favorable rainfall bumper rabi crops are grown in these areas. The major portion of the region is also covered by medium black soil. The medium and deep black soil in the region is rich in plant nutrients and can support good kharif and rabi crops like jowar, bajra, cotton, groundnut, wheat, tur, gram, safflower, sunflower, and soyabean. The deep black soil covers about 12.95 per cent portion of the Marathwada region. The medium black soil covers about 64.79 per cent and the coarse and shallow soil covers about 22.26 per cent portion of the Marathwada region in the year 2001. The average annual rainfall is not uniform in all the districts of Marathwada region. Marathwada region receives an average annual rainfall of 827 mms.

According to the population census of 2001 the total population of the region is 155.89 lakhs and about 75.41 per cent of total population lives in rural areas and 24.59 per cent in urban areas.

In Marathwada region out of the total geographical area of 64,43,400 hectares about 3.51 per cent (2,26,100 hectares) has been classified as forest area under land utilization statistics of 1999-2000 data maintained by the Agriculture Department. The major portion of the total geographical area i.e. 71.85 per cent is the net area sown. The average size of land holdings in the Marathwada region has declined from 6.18 hectares in 1970-71 to 2.39 hectares in 1990-91. The percentage of operational holdings has increased from 17.71 per cent in the year 1970-71 to 23.66 per cent in the year 1990-91. The cropping pattern in almost all the districts is dominated by food crops.

Major crops grown in the region are the kharif jowar, bajra, tur, cotton, sugarcane, groundnut, wheat, rabi jowar, gram, safflower, sunflower, sesamum and newly introduced crop soyabean alongwith other crops are also grown in Marathwada region.

In the Marathwada region, area under foodgrains was 72.43 per cent in the year 1981-82 which has decreased to 64.12 per cent in the year 1999-2000. The percentage area under cotton in all the districts in Marathwada region has continuously increased during the period 1981-2000. The percentage area under sugarcane and oilseed crops has slightly increased from 1.27 per cent, 11.47 per cent in 1981-82 to 2.08 per cent, 12.78 per cent in the year 1999-2000 respectively. The percentage area under groundnut has continuously declined from 2.23 per cent in 1981-82 to 1.50 per cent in 1999-2000 in Marathwada region. The percentage area under safflower has declined from 4.93 per cent in 1981-82 to 3.43 per cent in 1999-2000. But the percentage area under sunflower and soyabean has increased from 0.69 per cent, 0.01 per cent in 1981-82 to 5.13 per cent , 0.52 per cent in the year 1999-2000 respectively. In short the percentage area under oilseeds has continuously increased during the period 1981-2000. Lastly, the percentage of total area irrigated to total cropped area in the Marathwada region has continuously increased. It has increased from 10.59 per cent in the year 1981-82 to 14.33 per cent in the year 1999-2000.

Marathwada is one of the backward regions of the Maharashtra state. The rate of agricultural mechanization in the region is slow due to low crop intensity along with unawareness of advanced technology among the farmers and also the knowledge towards the use of unconventional energy sources. Marathwada is industrially backward because the use of raw materials locally available are not optimally used for industrial purpose. Low literacy rate is also one of the reasons of backwardness of Marathwada region. Even after many planning years neglect of authorities to implement better growth plans and lack of attention towards the socio-economic problems of the region still persist. Hence, more and more encouragement in the direction of opening agro-based and bio-technological units are required in the Marathwada region which will promote agricultural, industrial development and economy of the Marathwada region.

FINDINGS

The main findings of the present study are summarized as below:

(1) The study of growth performance of groundnut crop revealed that the groundnut production has decreased even though there was a positive growth in the yield of this crop during the overall period in Marathwada. This was due to negative growth in the area under this crop in Marathwada region as a whole. But the districtwise picture revealed that the growth performance of groundnut during the overall period has also been disappointing in Nanded, Latur, Osmanabad and Jalna districts and production have substantially increased in Parbhani, Beed and Nanded districts. The growth of area and production of groundnut have increased in Beed district only and decreased in Osmanabad, Jalna, Latur and Nanded districts. Area, production and yield growth rates of groundnut are positive in Beed district and negative in Latur district. Growth rates of yield of groundnut in most of the districts are positive and negative in Osmanabad district only.

(2) In case of safflower also the production has shown a negative growth even though there was a positive growth in the yield of this crop. This was due to negative growth in the area under this crop in Marathwada region as a whole, during the overall period. Safflower production performance was better in Nanded district and followed by Latur district but in rest of the districts it was found to be negative due to negative growth rate of yield of safflower. Area, production and yield growth rates were negative in case of Beed, Jalna and Aurangabad districts during the overall period. In most of the districts area,

production and yield growth rates of safflower were positive in sub-period I but during the sub-period II, it was negative alongwith Marathwada region as a whole.

(3) Sunflower area and production growth performance was better in all the districts of Marathwada region during the sub-period I and in overall period also. Yield growth rates were also positive in most of the districts except in Jalna district, even though there was a negative growth in the yield of this crop in Marathwada region as a whole. It means that the Marathwada region itself proved to be important producer of sunflower crop. Comparatively sunflower production growth was better in Parbhani district due to the higher area growth rate and production growth was lower in Latur district due to the lower growth rate of area of sunflower during the overall period. It is clear that the cause of fluctuations in sunflower production is the change in growth rate of area of that crop in Marathwada region during the overall period.

(4) Production growth performance of soyabean was better with the zero growth rate of yield. This was due to the expansion of area under the soyabean crop in Marathwada region as a whole, during the overall period. But the district level picture showed that the growth performance of production and area of this crop was better in most of the districts except Beed and Aurangabad districts. Production growth of this crop was highly satisfactory in Parbhani district and followed by Nanded district. Yield growth is negligible and area growth is remarkable in case of soyabean during overall period in Osmanabad district. Production performance was very poor due to negative area growth rate in Beed district.

(5) The study of production growth of oilseed crops has indicated that the production growth of sunflower crop was better due to the expansion of area under

that crop in Marathwada region during the overall period. As well as the yield growth was also satisfactory. The production growth performance of soyabean in most of the districts in Marathwada region was satisfactory during the overall period. The yield growth of groundnut crop in most of the districts in Marathwada region was found to be positive during the overall period. The area growth of safflower was positive in most of the districts but production growth was not positive in most of the districts of the Marathwada region. The area growth performance of groundnut was very poor in most of the districts except Beed district during the overall period in Marathwada region. We concluded that Marathwada region is a producer of sunflower crop and followed by soyabean during the overall period.

(6) The farmers' response to relative price of groundnut is found to be positive and insignificant in case of Jalna, Beed, Parbhani and Nanded districts, and negative and insignificant in case of Aurangabad, Osmanabad and Latur districts and for Marathwada region as a whole for sub-period I. And for sub-period II, it is found positive for Aurangabad, Beed and Osmanabad districts and negative for Jalna, Latur, Parbhani and Nanded districts but for overall period the farmers response to relative price of groundnut is found to be positive but insignificant for Aurangabad, Jalna, Beed and Osmanabad districts and Marathwada region as a whole, and negative and insignificant in rest of the districts. The relative yield factor has found to be positive but insignificant in Aurangabad, Beed, Osmanabad, Latur, Parbhani and Nanded districts and Marathwada region as a whole, but it is negative in case of Jalna district only. The gross irrigated area has indicated a negative impact on groundnut acreage in most of the districts and Marathwada region as a whole during the overall period. The sowing period rainfall has significantly affected the area under

groundnut in Parbhani district and Marathwada region as a whole. The lagged acreage variable has positive and significant impact on acreage under groundnut in all the districts and Marathwada region as a whole. The coefficient of acreage adjustment for groundnut crop is found highest for Jalna district and lowest for Parbhani district. The short run acreage elasticity with respect to price was found positive for Marathwada region during the overall period and negative and low in case of Aurangabad district during sub-period I. The highest short-run acreage elasticity with respect to price is found in Jalna district for sub-period I. Long-run acreage elasticity is also the highest in Jalna district during sub-period II.

(7) The impact of relative price of safflower exerted positive but insignificant impact on safflower acreage in case of most of the districts of the region and region as a whole. In Beed, Osmanabad and Parbhani districts it is found negative and insignificant during for the overall period. The relative yield variable has shown positive and significant impact in Jalna, Beed, Osmanabad and Parbhani districts and Marathwada region as a whole during the overall period. The gross irrigated area has shown negative and insignificant impact on acreage under safflower in most of the districts and Marathwada region as a whole. But it is found positive and significant in Osmanabad district only. The rainfall received during sowing period has shown positive influence on acreage under safflower in Marathwada region. But the coefficient of rainfall variable is found to be significant in Aurangabad, Beed, Latur, Osmanabad and Parbhani districts and for Marathwada region as a whole during the second sub-period. The impact of lagged acreage is found to be positive for safflower crop in all the districts and for Marathwada region as a whole during the overall period. The coefficient of adjustment with respect to safflower crop for Marathwada region is found to be

highest during the sub-period II. But it is lowest and negative in Parbhani district for sub-period II and highest in Beed district during the same period. The short run acreage elasticity with respect to price for Marathwada region is positive but it is negative and very low in case of Osmanabad district during the overall period. In Parbhani district it is found to be highest during the sub-period II as well as the long run acreage elasticity with respect to price is highest in Parbhani district during the sub-period I.

(8) The effect of relative price of sunflower is found to be positive but insignificant in Jalna, Osmanabad, Latur and Parbhani districts and Marathwada region as a whole. In case of Aurangabad and Beed districts it is found positive and significant during the overall period. In case of Nanded district effect of this variable is negative during all the periods. The effect of this variable is found positive and significant in most of the districts during sub-periods I and II. The relative yield factor has exerted a positive impact on acreage in most of the districts and Marathwada region as a whole. In case of Aurangabad, Parbhani and Nanded districts it is found negative, only in case of Beed district it is found positive and significant during the overall period. The gross irrigated area has indicated negative influence on sunflower acreage in all the districts of Marathwada region during the overall period. The rainfall received during the sowing period has insignificant impact in most of the districts of Marathwada and Marathwada region as a whole during the overall period. But the sowing period rainfall has significantly affected the area under sunflower in case of Osmanabad district and it has negative effect in Nanded district. Lagged acreage of sunflower has positive and significant influence on acreage under sunflower in all the districts and Marathwada region as a whole. The coefficient of acreage adjustment for sunflower is found highest in

Beed district and the lowest in Latur district during the overall period. The short-run acreage elasticity with respect to price is found to be -0.0082 for Marathwada region as a whole during the overall period. But it is high in case of Aurangabad district and lowest and negative in Beed district during the overall period. Long run acreage elasticity with respect to price for Aurangabad district is found to be the highest.

(9) The farmers response to the relative price of soyabean is found to be positive and insignificant in Marathwada region. The impact of yield is found to be negative and insignificant. The gross irrigated area has exerted negative impact on soyabean in the region. The rainfall received during sowing period has positively affected the area under soyabean. Lagged acreage has exerted positive and significant influence on acreage under soyabean in Marathwada region. The coefficient of acreage adjustment for soyabean is found to be 0.294. The short run acreage elasticity with respect to price is -0.0340 and long-run acreage elasticity with respect to price is -0.1156.

POLICY MEASURES

Now it is well accepted that oilseed production cannot sustain in the assured irrigated states, regions and districts. These crops cannot compete with the high value cereals. Therefore substantial contribution to the oilseed production at the national level has came from those states, regions and districts which do not have a competitive advantage in the production of cereals. The findings of the present study listed above makes us to state the policy measures which may prove useful for the sustainable growth of the oilseed crop economy of the Marathwada region.

(1) The production of groundnut crop decreased even though there was a positive growth in yield of this crop. This was due to negative growth in area of this crop in most of the districts of the region. This implies

concentrated efforts in developing new varieties of groundnut crop whose yield potential is stable across different agro-climatic districts. The low production and/or productivity of groundnut crop is mainly on account of low application of new inputs like HYV seeds, fertilizers, pesticides, growing crop in rainfed areas; pushing this crops to marginal lands,. Domination of high value crops in the cropping pattern and poor management of the farmers in the region.

(2) In our study we have come to the conclusion that safflower production in Marathwada region as a whole and in most of the districts has shown a declining trend during overall period as well as in sub-period II also. The findings of our study indicate that most of the districts of the region accounted negative growth rate of yield of safflower crop,while yield growth was positive in Nanded district.To increase the yield of safflower crop we suggest the improved and HYV seeds, fertilizers and pesticides are to be provided to the producers of safflower crop in Marathwada region. Then yield of safflower crop will be increased. When yield increases the production of this crop would increase in all the districts of the Marathwada region. But special efforts are to be made in Jalna district because production and yield growth rates of safflower were very low for this district..

(3) The present study has shown that the area and production growth rates of sunflower were positive and yield growth rate was negative in Marathwada region as a whole. But district level picture revealed that in most of the districts area, production and yield growth rates of sunflower were positive in Marathwada region during the overall period. Yield of sunflower crop has recorded negative growth in Nanded district only. During the same period the growth of sunflower production is highest in Parbhani district followed by Aurangabad and Nanded districts. This helps us to

suggest that efforts are needed to boost the yield of sunflower in Marathwada region as a whole and similar efforts should also be undertaken in Jalna district such as education and training regarding cultivation of sunflower is also essential for raising yield. Irrigation facilities, HYV seeds, fertilizers, pesticides and insecticides and credit facilities should be provided to the producers of sunflower crop in the region.

(4) The study of growth performance of soyabean crop revealed that the soyabean production has substantially increased even though there was a zero growth rate in the yield of this crop during the overall period in Marathwada region. The area, production and yield growth rates were positive even though soyabean is a newly introduced oilseed crops in Marathwada region. To increase the production of soyabean, the area under this crop must be increased especially in Parbhani, Beed and Aurangabad districts. To increase the area under this crop, creation of awareness in production of this crop is essential and improved seeds, new technologies of production, fertilizers, pesticides and insecticides and market facilities must be provided to the producer of this crop in Marathwada region.

(5) The lagged acreage and sowing period rainfall are found to be the most significant factors affecting acreage under groundnut in Marathwada region as a whole and to some extent relative price and relative yield are found to have exerted positive impact on acreage of this crop. The measures which ensure higher gross return per hectare from groundnut needs to be adopted. The gross return depends upon the level of price and yield. All the yield raising measures like fertilizers, HYV seeds, pesticides, credit, new methods of cultivation and new technology should be adopted along with appropriate price policy to increase groundnut production in Marathwada region. The impact of lagged acreage variable is positive and

statistically significant in most of the districts and Marathwada region as a whole. This indicates that the producer of groundnut was not market oriented . Price incentives be provided to the groundnut producer so that they would begin to produce this crop for market.

(6) The farmers in most of the districts in Marathwada region have indicated positive response to relative price of safflower. This suggests that this crop is grown for family consumption as well as for marketing purpose. The price policy should be made favourable to this crop so that farmers growing safflower would respond more significantly and production will increas. The relative yield variable has also found positive and significant impact in Jalna, Beed, Osmanabad and Nanded districts and Marathwada region as a whole. In this situation yield raising measures should also be adopted in case of safflower crop in Marathwada region. Sowing period rainfall and lagged acreage are found to be the most significant factors affecting acreage under this crop in most of the districts of the region which suggests the need of yield oriented policy for this crop. The yield raising policy measures, incorporating timely and adequate supply of inputs like HYVs of seeds, chemical fertilizers, bio-fertilizers, pesticides, insecticides and credit facilities at reasonable interest rate along with remunerative prices to the farmers should be adopted.

(7) The sowing period rainfall, relative yield, relative price and lagged acreage have affected positively and significantly the acreage under sunflower in most of the districts of Marathwada region. Acreage under this crop can also be increased by adopting yield-raising policy measures and by ensuring remunerative price to sunflower growers. Such measures are necessary for raising the production of sunflower in most of the districts of Marathwada region. The impact of lagged acreage is positive and statistically significant in all the districts of Marathwada region during the overall

period. This shows that the farmers maintain their acreage under sunflower crop to fulfill their family needs of consumption. Price incentives be provided to the sunflower growers so that they would begin to produce sunflower crop for the market and production of sunflower would become market oriented.

(8) Soyabean is a newly introduced oilseed crop in Marathwada region. It is an important cash crop grown in this region . The acreage under this crop is positively affected by the relative price, lagged acreage and sowing period rainfall. The emphasis needs to be given on yield raising measures along with suitable price and marketing policies in case of soyabean to raise the area and production of this crop in Marathwada region. Price incentives be provided to the soyabean growers so that they would begin to produce this crop for the market.

(9) The findings of our study suggests that government should provide adequate irrigation facilities to achieve higher level of oilseeds production. To increase oilseeds productivity further, much needs to be done especially in respect of full utilization of water. In order to increase full utilization of water, government should provide adequate inputs such as fertilizers, improved and HYV seeds, pesticides, insecticides, new technology of cultivation, adequate credit at reasonable rate, etc. to the farmer at a proper time and adequate knowledge about water management should be given to the farmers through extension services and training. Government should also provide to the farmers marketing and storage facilities and reasonable prices on the basis of cost of production for oilseed crops. Besides, government should develop some new methods of dry farming. Drought resistant varieties should be invented and use of these varieties should be made popular among the farmers. Marathwada Agricultural University and Agricultural colleges in Marathwada region can do a lot in this respect by

undertaking research in dry farming techniques for sustainable development of oilseed economy of the Marathwada region. Water management can be made by establishing water distributing cooperative societies in the region. Soil and water conservation measures must be adopted by the Government to overcome the problems of dry oilseed farming in the Marathwada region.

(10) The oilseed crop economy of Marathwada may develop at a rapid rate if the importance is given to the development of non-traditional oilseeds like soyabean and palm along with the traditional oilseed crops in the region. There is a further scope to develop the oilseed economy by giving emphasis on tree oriented oilseeds. This type of efforts not only will help in hiking oilseed production but will help in having afforestation to maintain ecological balance which is the need of the day.

All these above measures should be adopted simultaneously to achieve higher level of oilseeds production in the Marathwada region.

Bibliography

Agrawal, P. C. (1990), *Oil Seeds in India: Perspectives for 2001 A.D.*, Oxford & IBH Publishing Company Pvt. Ltd., New Delhi.

Acharya, K. T. (1990), *Oil Seeds and Oil Milling in India: A Cultural and Historical Surgery*, Oxford & IBH Publishing Company Pvt. Ltd., New Delhi.

Behraman, J. R. (1968), *Supply Response in Underdeveloped Agriculture: A Case Study of Four Major Annual Crops in Thailand*, North Holland Publishing Company, Amsterdam.

Bhagat, L. N. (1989), *Supply Response in Backward Agriculture–An Econometric Study of Chotanagpur Region (Bihar)*, Concept Publishing Company, New Delhi.

Bhise, V. B. (1988), Price Policy for Foodgrains in India in G.S. Kalyankar edited *'Indian Economy'*, Marathwada University, Aurangabad.

Bapna, S. L. (1980), *Aggregate Supply Response of Crops in Developing Region*, Sultan Chand and Sons, New Delhi.

Dharam Narain (1965), *Impact of Price Movement on Areas under Selected Crops in India, 1900-1939*, Cambridge University Press.

Gupta, R. P. (1973), *Agricultural Prices in Backward Economy*, National Publishing House, Delhi.

Green Maurice B. (1978), *Eating Oil: Energy Use in Food Production*, West View Press, Boulder, Colorado, U.S.A.

John, P. V. (1965), *Some Aspects of the Structure of Indian Agricultural Economy, 1947-48 to 1961-62*, Asia Publishing House, New Delhi.

Kumar, D. and Rai, M. (1992), *Advances in Oilseeds Research*, Scientific Publishers, Jodhpur.

Krishna, Raj (1967), Price Policy and Economic Development in H.M. Southworth and B.F. Johnston (Edt), *Agricultural Development and Economic Growth,* Cornell University Press, Ithaca, New York.

Kaur, Rajbans (1975), *Agricultural Price Policy in Economic Development,* Kalyani Publishers, Delhi.

Mahajan, V. S. (1985), *Growth of Oilseeds and Allied Industries in India,* Deep and Deep Publication, New Delhi.

Nerlove, Marc (1958), *Dynamics of Supply: Estimation of Farmers Response to Price,* John Hopkins Press, Baltimore.

Sawant Shashikala (1978), *Supply Behaviour in Agriculture-An Econometric Analysis,* Himalaya Publishing House, Bombay.

Sulbha Brahme, Kumud Pore, S. H. Pore (1975), *Regional Planning: A Case Study of Marathwada Region,* Gokhale Institute of Politics and Economics, Pune.

Satish Chandra (1985), *Agricultural Price Policy in India,* Chug Publication, Allahabad.

Tyagi, D. S. (1974), *Farmers Response to Agricultural Price in India: A Study in Decision Making,* Heritage Publishers, Delhi.

Uma Kapila (1982), *Acreage Response of Groundnut-A State Level Study, Oil Seeds Economy of India,* Agricultural Publishing Academy, New Delhi.

Acharya, S. S. and Agrawal, N. L. (1974), Factors Affecting Supply and Growth Rates of Commercial Crops in Rajasthan, *Indian Journal of Agricultural Economics,* Vol. XXIX, No. 3.

Acharya, S. S. and Bhatiya, S. (1974), Acreage Response to Price, Yield, Rainfall Changes in Rajasthan, *Agricultural Situation in India,* Vol. XXVIII, No. 4.

Askari Hossein and Cummings, T. T. (1974), Output Price Response in Agriculture: An Evaluation. *Indian Journal of Agricultural Economics,* Vol. XXXIX, No. 2.

Bhagat, L. N. (1985), Determinants of Wheat-Acreage Fluctuations in Bihar: A Study of Temporal Spatial Variations, *Indian Journal of Agricultural Economics,* Vol. XL, No. 2, April-June.

Chandrabanu, P. and Thomas, E. K. (1991), Supply Behaviour of Sesamum and Groundnut in Kerala, *Agricultural Situation in India,* Vol. XLVI, No. 5, August.

Dixit P. K., Hiremath, K. L. and Rajvir Singh (1998), Production Behaviour of Groundnut Farmers in Karnataka: A Dynamic

Supply Response Analysis. *Indian Journal of Agricultural Economics,* Vol. 53, No. 2.

Jhala, M. L. (1979), Farmers Response to Economic Incentives: An Analysis of Inter-Regional Groundnut Supply Response in India, *Indian Journal of Agricultural Economics,* Vol. XXXIV, No. 1.

Kamala Devi, P. and Rajgopalan, R. (1965), Price and Acreage Response–A Case Study of Groundnut Crop in North Arcot District, Madras State, *Indian Journal of Agricultural Economics,* Vol. XXI, No. 1.

Kaul, J. L. and Sidhu, D. S. (1971), Acreage Response to Prices for Major Crops in Punjab–An Econometric Study, *Indian Journal of Agricultural Economics,* Vol. XXVI, No. 4.

Krishna Raj (1963), Farm Supply Response in India–Pakistan: A Case Study of Punjab Region, *The Economic Journal,* Vol. LXXIII, No. 29, September.

Krishna Raj (1963), Rapporteur's Report on Economics of the Cropping Pattern, *Indian Journal of Agricultural Economics,* Vol. XVIII, No. 1, January-March.

Madhavan, M. C. (1972), Acreage Response of Indian Farmers–A Case Study of Tamil Nadu, *Indian Journal of Agricultural Economics,* Vol. XXVII, No. 1.

Misra, V. N. and R. Radhakrishnan (1971), Economic Models for Acreage Response–A Study of Groundnut and Bajra in Gujrath, *Anvesak,* Vol. 1, No. 2.

Mistra, B. and Singha, S. P. (1958), Agriculture and its Terms of Trade with Special Reference to India, *Indian Journal of Agricullural Economics.*

Mohapatra, B. K. (1972), The Impact of Price Changes on Areas Under Selected Crops in Orissa, *Asian Economic Review,* Vol. 14.

Narayana, N. S. S. and Parikh, K. S. (1979), Agricultural Planning and Policy in Draft Sixth Plan: Will Farmers Fulfill Planners Expectation? *Economic and Political Weekly,* Vol. XIV, No. 31 and 32, Special Number, August.

Nerlove Marc (1958), Distributed Lags and Estimation of Long Run Supply and Demand Elasticities–Theoretical Consideration, *Journal of Farm Economics,* Vol. XL, No. 2, February.

Padwal, R. N. (1954), Production and Price Trends of Major Oilseeds in India, *Indian Journal of Agricultural Economics,* 9(56).

Pal, S. P. (1975), Supply Response and Optimal Agricultural Policy in a Planned Underdeveloped Economy, *Indian Journal of Agricultural Economics,* Vol. XXX, No. 2, April-June.

Parikh, A. (1972), Market Responsiveness of Peasant Cultivators: Some Evidence from Prewar India, *Journal of Development Studies,* No. 8.

Ravikumar, K. N., K. Sree Lakshmi and Raju, V. T. (2000), Price Response Analysis of Agricultural Commodities in the Selected Regulated Markets of Andhra Pradesh, *Agricultural Situation in India,* Vol. LVI, No. 12, March.

Reddy, J. M. (1970), Estimation of Farmers Supply-A Case Study of Groundnut, *Indian Journal of Agricultural Economics,* Vol. XXV, No. 4.

Sahay, K. K. (1971), Acreage Response of Groundnut in Major Groundnut Zones: A Case Study in Dynamic Supply, *Agricultural Situation in India,* Vol. XXVI, No. 7.

Shakuntala Gupta (2000), Acreage Response of Major Oilseed Crops in Punjab, *Agricultural Situation in India,* Vol. LVI, No. 12, March.

Shreeram Murthy C. and Eshwari Prasad (1990), Supply Response Analysis of Sesamum in Andhra Pradesh, *Agricultural Situation in India,* Vol. XLIV, No. 11.

Sidhu, D. S. and Prakash Mehta (1975), Response of Farmers to Prices, A Case Study of Potato and Groundnut in Punjab, *Economic Affairs,* Vol. XXIX, No. 6.

Sidhu, J. S. and Sidhu R. S. (1988), Growth and Area Response of Commercial Crops in Punjab : Someo Policy Issues, *Indian Journal of Agricultural Economics,* Vol. XLIII, No. 3.

Singh, O. P. (1998), Growth and Response of Oilseeds in Uttar Pradesh, *Agricultural Situation in India,* Vol. LV, No. 1.

Singh, R. P., Singh Parmatma and Raj, K. N. (1973), Acreage Response to Rainfall, New Farm Technology and Price in Haryana, *Indian Journal of Agricultural Economics,* Vol. 54, No. 4.

Tripathy, S. and Srinivasa Gowda, M. V. (1993), An Analysis of Growth, Instability and Area Response of Groundnut in Orissa, *Indian Journal of Agricultural Economics,* Vol. 48, No. 3.

Hedgire, D. N. (1992), A Study of Production, Marketing and Processing of Major Edible Oil Seeds in Marathwada-An Economic Analysis, Unpublished Ph.D. Thesis, Marathwada Agricultural University, Parbhani.

Kawarkhe, P. K. (2001), Economic Efficiency of Sugar Industry in Marathwada Region of Maharashtra, Unpublished Ph.D. Thesis, Marathwada Agricultural University, Parbhani.

Patil, K. K. (1997), Impact of Price and Non-Price Factors on Sugarcane Acreage in Marathwada, Unpublished Ph.D. Thesis, Dr. Babasaheb Ambedkar Marathwada University, Aurangabad, October.

Panpatte, B. P. (1980), Some Structural Aspects of Agricultural Development in Marathwada, Unpublished Ph.D. Thesis, Marathwada University, Aurangabad, June.

Rajgopalan, V. (1967), Supply Response for Irrigated Crops in Madras State, India, Ph.D. Dissertation, University of Tennessee.

Garule, J. H. (1984), Acreage Response of Kharif Jowar in Parbhani District of Maharashtra State, M.Sc.(Agri), Thesis, Marathwada Agricultural University, Parbhani.

Sable, M. S. (1991), Supply Response of Some Agricultural Commodities in Marathwada, Unpublished Ph.D. Thesis, Marathwada University, Aurangabad, August.

Kanhere, R. B. (1997), Growth and Instability in Oilseeds Crop Economy of Maharashtra, Unpublished Ph.D. Thesis, Dr. Babasaheb Ambedkar Marathwada University, Aurangabad, June.

Director: Government Printing and Stationery, Maharashtra State, Bombay, Season and Crop Reports, Maharashtra State, 1979-80 to 1998-99.

Government of Maharashtra: Agriculture and Cooperative Department, Epitome of Agriculture in Maharashtra, Part I and II, 1984-85 to 2001-2002.

District Statistical Officer: Socio Economic Review and District Statistical Abstracts of Aurangabad, Jalna, Beed, Osmanabad, Latur, Parbhani and Nanded Districts, for the Years 1979-80 to 2001-2002.

Director: Government Printing and Stationery, Maharashtra State, Bombay, Census of Maharashtra 1981–Series and 1991 Series–14 and Population Census, 2001.

Government of Maharashtra, Human Development Report, 2002.

Director of Economics and Statistics, Regional Office, Aurangabad, At a Glance Aurangabad Region, 2002.

Marathwada at a Glance: A Case Study Conducted by Marathwada Agricultural University, Parbhani, 1996.

Director: A Report of National Seminar on Stress Management in Oil Seeds for Attaining Self-Reliance in Vegetable Oils, Organized by Director of Oilseed Research (ICAR), Rajendranagar, Hyderabad, January 28-30, 2003.